COUNTRYSIDES AT RISK: THE POLITICAL GEOGRAPHY OF SUSTAINABLE AGRICULTURE

POLICY ESSAY NO. 16

COUNTRYSIDES AT RISK:

THE POLITICAL GEOGRAPHY OF SUSTAINABLE AGRICULTURE

Robert L. Paarlberg

Overseas Development Council
Washington, DC

Library of Congress Cataloging-in-Publication Data

Paarlberg, Robert L.
Countrysides at risk: the political geography of sustainable agriculture/Robert L. Paarlberg

Policy Eassy No. 16
Includes bibliographic references.
1. Agriculture—Environmental aspects. 2. Agricultural ecology. 3. Sustainable agriculture. 4. Agriculture—Environmental aspects—Developing countries. 5. Agricultural ecology—Developing countries. 6. Sustainable agriculture—Developing countries. I. Title. II. Series.

S589.75.P33 1994 333.7'137'091724—dc20 94-43973 CIP

ISBN: 1-56517-021-0

Printed in the United States of America.

Director of Publications: Christine E. Contee
Publications Editor: Jacqueline Edlund-Braun
Edited by Elizabeth Reed Forsyth
Cover design: Snoreck Design Group
Book design: Tim Kenney Design Partners, Inc.

Contents

Foreword vii

Executive Summary 1

Part I. Introduction 11

Countrysides and Food Supplies at Risk 12

Part II. Asia 17

First- and Second-Generation Environmental Damage from Farming 18

First-Generation Problems 18

Second-Generation Problems 28

Can Growth of Yield in Irrigated Rice Farming Be Sustained? 31

Summary Prescriptions 33

Part III. Africa 37

First-Generation Environmental Problems as a Constraint on Future Production of Food 38

Natural Disadvantages 42

Difficult Demographics 43

Poor Governance 45

Examples of Intensification 48

Summary Prescriptions 50

Part IV. Latin America 53

Dualistic Farming, Dualistic Resource Abuse 54

Historical Foundations of Dualism 54

Mexico: Limited Environmental Gains from Structural Adjustment 57

Central America: Limited Gains from Technical Change 61

Summary Prescriptions 65

Part V. The Role of External Actors 67

International Assistance 68

U.S. Support for Agricultural Research 68

World Bank Support for Sustainable Agriculture 71

Nongovernmental Organizations as Environmental Advocates 75

Part VI. Conclusion 81

Resource Abuse as Power Abuse 82

Notes 86

About the Author 96

About the ODC 97

Board of Directors 98

Foreword

Despite massive progress in food production in recent years, some 700 million people in developing countries go hungry each year. Population growth is creating additional, unprecedented demand for food production. As Robert Paarlberg points out in this *Policy Essay*, farmers in poor countries will have to produce more food over the first two decades of the next century than they have produced over the entire 10,000 years since sedentary farming first began.

Increasing food output today almost always implies increasing pressure on soil and water resources. Most developing countries have already cultivated their best land, and increasing food output requires either cultivation of ecologically vulnerable environments—tropical forests, dryland and hilly areas, and the fringes of great urban centers—or increasing the use of agricultural chemicals and other inputs.

Environmentally sustainable farming is clearly part of the solution. Yet there is no consensus on what such farming policies and practices should actually look like. Production-oriented agriculturalists argue that environmental protection—especially protection of forests and topsoil—can be advanced through modern farming techniques that require heavy application of water and fertilizer. Environmental advocates, by contrast, associate such high-input farming with chemical pollution, a faster exhaustion of water supplies, and a dangerous loss of biodiversity.

In part due to a lack of clear technical direction, international agricultural assistance is declining despite the growing food needs of developing countries. World Bank lending for all agricultural and rural development projects has been shrinking in real terms for more than a decade, falling from 30 percent of total Bank lending in 1978 to 14 percent in 1993.

Robert Paarlberg, a professor at both Wellesley College and Harvard University, is one of this country's experts on agricultural policy, environmental policy, and the linkage between the two. In this *Policy Essay* he urges that policymakers adopt a set of geographically differentiated policies. Different regions of the world require different technical packages to boost food production while protecting the environment. And, as Paarlberg usefully reminds us, in some regions, technical changes, no matter how appropriate, will not occur until more fundamental political and social reforms are implemented to address the underlying problems of power abuse that negatively affect people and their use of natural resources.

Countrysides at Risk is the latest in ODC's body of work on the relationship between the environment and development. *Poverty, Natural Resources, and Public Policy in Central America* by Sheldon Annis and contributors offers analyses and concrete proposals on strategies to reduce poverty and protect the environment in the region, including among other issues the environmental implications of promoting nontraditional agricultural exports and the impact of changed land tax policies on the poor. *Environment and the Poor: Development Strategies for a Common Agenda* by H. Jeffrey Leonard and contributors focuses specifically on the huge and still rapidly growing numbers of very poor people (perhaps as many as six out of ten of all the developing world's poor) who are being forced onto marginal lands with disastrous implications for the environment and their own well-being. All of these works make it clear that there are no short cuts in the effort to implement those measures that can reduce poverty, meet food needs, stabilize rapidly expanding populations, and safeguard the environment and natural resources on which those countries so greatly depend.

Countrysides at Risk is part of ODC's core program of policy analysis, which is supported by The Ford Foundation and The Rockefeller Foundation.

John W. Sewell
President
December 1994

Acknowledgments

The international research needed to prepare this essay was generously supported by The Ford Foundation. Other institutional debts of gratitude are owed to the Overseas Development Council, the International Food Policy Research Institute, Winrock International, the Food and Agriculture Organization of the United Nations, the International Crops Research Institute for the Semi-Arid Tropics, the International Institute of Tropical Agriculture, the World Resources Institute, the Harvard Center for International Affairs, and Wellesley College. Valuable professional assistance was received from Nikos Alexandratos, Pierre Antoine, Sandra S. Batie, Robert J. Berg, Richard E. Bissell, Kevin M. Cleaver, Christine Contee, Ralph W. Cummings Jr., Dana Dalrymple, Christopher L. Delgado, Paul Faeth, Anthony Gambino, Neva R. Goodwin, Merilee S. Grindle, Catherine Gwin, Peter M. Haas, Peter B. R. Hazell, Robert W. Herdt, Idriss Jazairy, Ernst Lutz, Michel Petit, Patti L. Petesch, Per Pinstrup-Andersen, Vernon W. Ruttan, David Seckler, Dunstan S. Spencer, Lori Ann Thrupp, Steve Vosti, Jennifer A. Widner, and Montague Yudelman. Special personal thanks are owed to Patty Allison, Robert O. Blake, Walter Coward, Jr., Lowell Hardin, Robert D. Havener, Hans R. Herren, Earl Kellogg, John Kerr, Michael Lipton, Anne Meixsell, Don Paarlberg, Eric Rusten, John W. Sewell, and T. Kelley White. My readers and editors at the Overseas Development Council, together with my seminar students at Wellesley, provided candid reactions that helped sharpen the final product. I dedicate this work to my wife, Marianne Perlak.

Executive Summary

Farming is a threat to the natural environment in rich as well as poor countries, but the human stakes are now much higher in the developing world, where food needs are acute and growing rapidly. Because of population growth, the developing world is being asked to feed 88 million *additional* people every year. How can this production challenge be met if environmentally destructive farming practices continue?

In much of Africa, where crop yields will have to increase, the "mining" of soil nutrients is now helping to push average crop yields into decline. In much of South Asia, old irrigated lands are becoming saline and waterlogged and are going out of use almost as fast as new irrigated lands are coming into production. From Honduras to Java, soils are washing away on newly cleared sloping lands. In East Asia, South Asia, and Central America, the natural biological controls for crop pests are being poisoned with farm chemicals, even while the pests themselves are becoming more resistant to poison.

Worsening this crisis is a paralyzing technical debate between agriculturalists and environmentalists over what environmentally sustainable farming should look like. Production-oriented agriculturalists argue that environmental protection—especially of forests and topsoil—can be advanced best through modern, input-intensive farming. Environmental advocates, by contrast, associate high-input farming with chemical pollution, faster exhaustion of water supplies, and dangerous loss of biodiversity. They feel it is better to hold onto traditional farming techniques suited to local ecologies and the circumstances of ordinary resource-poor farmers.

These divergent positions have helped paralyze the international policy community. Bilateral and multilateral assistance organizations, not wishing to offend powerful environmental lobby groups, have become increasingly wary of sponsoring input-intensive farm modernization projects—one reason why international assistance to farming and farm research has recently faltered. External assistance to agriculture in low-income developing countries declined by 17 percent between 1980 and 1990. Yet the number of people needing food in the developing world grows every year, while the quality of their farm resource base continues to degrade.

How can this paralyzing policy deadlock be broken? Paying more attention to geography and politics is a start. In some regions of the developing world, agriculturalists are right to argue for more use of purchased inputs. In other regions, environmentalists are right to argue for less input use. In still other regions, neither group is entirely correct, because appropriate technical changes will not take place without more fundamental political and social change.

ASIA

■ RAPID, SUSTAINED, AND EQUITABLE economic growth sets much of Asia (especially East Asia) apart from both Africa and Latin America. In Asia, strong gains in food production made possible by science-based and input-intensive green revolution farming technologies have contributed significantly to this success. Particularly during the decades of the 1960s and 1970s, massive investments were made in crop research and irrigation infrastructure for production of food grains such as rice and wheat. Between 1965 and 1985, rice yields in Taiwan and Korea increased 53 and 132 percent, respectively. In India, irrigation and high fertilizer use in combination with newly available seed varieties made possible a rapid doubling of wheat production. The success of input-intensive production brought widely shared gains in nutrition, employment, and social welfare.

Agriculturalists argue that this intensification of production also helped feed people at the lowest possible cost to the rural environment. If India had attempted to use traditional low-yield farming techniques to double its wheat production, it would have had to plow up millions of hectares of additional cropland, resulting in further deforestation, added habitat destruction, and soil erosion. Environmentalists who criticize Asia's green revolution should acknowledge the region's need to boost total production and weigh the environmental damage that would have taken place if this had been attempted without a switch to input-intensive farming.

On the other hand, the environmentalists' preference for reduced input use is now fully justified in some of the more advanced Asian countries now undergoing rapid industrial development. The original switch to high-yield farming may have helped these countries ease a "first generation" of rural environmental problems (soil erosion, tree cutting, and habitat destruction), but it has now become associated with a dangerous "second generation" of problems (excess water and fertilizer use, inadequate nutrient and animal waste containment, loss of biodiversity, and excessive reliance on pesticides).

Asia's environmental gains from high-yield farming have also been limited in two other ways. First, highly productive farming has not yet been achieved in all regions of Asia, or by all farmers within those regions. Barriers to extending intensification are encountered in countries where farmers have insecure land tenure, in countries with large populations of landless rural laborers, in less-favored upland and dryland regions, and in countries where governments are not accountable to local farm communities.

Second, Asia's gains from intensification can be sustained in the critical decades just ahead only through continued public investment in research and infrastructure. Governments in the region and international donors have recently grown reluctant to provide these investments.

AFRICA

■ WHILE GREEN REVOLUTION FARMING has moved Asia toward second-generation rural environmental problems, most of Africa has yet to achieve a green revolution. As a consequence, Africa is still struggling to solve first-generation problems associated with expanding low-input farming or grazing onto more fragile lands or deeper into forest margins.

Africa's first-generation environmental problems are now so serious that they threaten to worsen the region's already grave crisis in food production. Africa is the only region in the world where per capita food production has been *declining*, a trend that reflects a general failure to boost crop yields.

Drought has contributed to this pattern of yield decline, but resource degradation is a factor as well. Nitrogen and phosphorus are being mined from African soils, and soil and water resources are being degraded by erosion, salinity, and sedimentation. If recent trends in the depletion of ligneous vegetation continue in the Sahel region, large numbers of cattle and people living there will find it increasingly difficult to survive the dry season. Deforestation is proceeding at a rate 30 times faster than reforestation, with roughly 5 million hectares lost every year, primarily as a result of clearance for agriculture.

Africa thus challenges the notion that rural environmental destruction results from input-intensive or science-based farm modernization. Africa has not yet embraced high-yield farming, yet it is still degrading its environment. Rather than suffering from excessive farm input use, Africa's rural environment suffers from too little: Fertilizer use in Africa is one-quarter the level of India and one-thirty-sixth the level of Japan; irrigation covers only 4 percent of cultivated area in Africa.

Environmentalists fearful of the second-generation problems that can accompany green revolution farming have been slow to accept the intensification imperative for Africa and to endorse the larger investments now needed in roads, irrigation, and non-traditional purchased inputs, such as inorganic fertilizer or hybrid seeds. Given the reality of rapid population growth, however, Africa has no practical alternative.

In taking the necessary steps to protect the environment by making farming more productive, Africans will face unique difficulties. Compared with East Asia, the continent is endowed with a more difficult climate and poorer natural resources for intensified farming; its demographic transition has been more sudden and more disruptive; and its political institutions are less stable, less accountable, and too heavily centralized.

Africa's governments tend to be urban-biased and nonaccountable to ordinary farming communities. Agriculture accounts directly for 80 percent of employment in Africa and for 50 to 90 percent of export earnings, but national governments typically devote less than 10 percent of their public spending to the farm sector. A recent inventory of 46 African governments found that only five of the smallest could be classified as genuine multi-party systems. Because so many African governments are non-democratic and reluctant to share power peacefully, they often invite armed challenge. As a consequence, Sub-Saharan Africa has recently had a per capita war fatality rate three times as high as that of the Middle East. Economic growth has not fared well amid the violence and dislocation, and neither has farm resource protection.

Africans and outsiders have yet to come to grips with the region's serious demographic and rural environmental crisis. To avoid a slow-motion Malthusian and ecological nightmare, larger investments must be made in rural roads, fertilizer, irrigation, extension, and research. Political institutions and policies must be reformed to give farmers and local farm communities greater participation and control, plus stronger incentives to make resource-conserving investments. Most African governments have been slow to respond to these massive challenges, yet outsiders have no grounds to complain so long as their own efforts remain so tentative and so small.

LATIN AMERICA

■ LATIN AMERICA'S AGRICULTURAL SECTOR exhibits a curious mix of both first-generation and second-generation environmental problems. Latin America is well endowed by nature, but not by its own social history. The highly inequitable control over both land and labor that was established at the time of the original European conquest survives today in the region's highly dualistic farm structures. A small elite of commercial farmers controls a disproportionate share of the region's high potential farmland and has tended to monopolize gains from modernization through privileged political access to subsidized water, machinery, and chemical inputs. Serious second-generation environmental damage on these commercial farms is a frequent result. Meanwhile, the majority of rural peasants farm smaller plots of poor or poorly watered land, with little intensification support from the government. These smaller farmers are inclined to do first-generation environmental damage.

The links between agricultural dualism and environmental destruction are so strong in Latin America that remedies based on macroeconomic policy reform or technical change alone have produced limited gains. Two case studies illustrate this point: the limited rural environmental gains that have come from market-oriented policy reforms in Mexico and from technical change (the introduction of high-value crop farming) in Central America.

In Mexico, a small elite owns the best irrigated lands and has profited over the years from generous (and environmentally damaging) government subsidies, while the vast majority of farmers are confined to lands with little water, poor soils, and disadvantageous terrain. Dualistic patterns of resource abuse have been the result.

Under the government of President Carlos Salinas de Gortari, Mexico undertook an ambitious and largely successful set of macroeconomic policy reforms, trade liberalization, and widespread market deregulation and privatization. The most important reform in the agricultural sector was a constitutional amendment that permitted private sale or rental of communal *ejido* lands. In theory, perhaps, this reform should have positive environmental effects, by allowing small peasant farmers to gain the incentive and collateral they need to finance resource conserving investments. Socially marginalized Mexican peasants, including many of indigenous background that have traditionally been victims of racial discrimination, need more than technocratic market reforms. If these reforms trigger widespread foreclosures, thus increasing the numbers of rural landless, disadvantageous environmental as well as social consequences will be the result.

Central America illustrates the limited value of trying to use technical change to resolve environmental problems rooted in rural social inequity. Movement toward higher value farm production can be good for the environment, as experience has shown in Asia and parts of Africa, where it allows farmers to substitute the use of labor and capital inputs for the use of forest lands, erodible soils, and scarce water. In Central America, however, the gains from farm modernization have usually been monopolized by only a small elite; ordinary peasants with limited legal protection have often found themselves displaced from subsistence farming once technical change has made their land more valuable (to elites with new options to engage in production and export of higher-value crops).

To contain both first- and second-generation rural environmental damage in Latin America (and also urban environmental damage, linked to excessive farm-to-town migration), farmers in the countryside who are currently weak and dispossessed will have to become beneficiaries of genuine social and political change, not just policy or technical change.

THE ROLE OF EXTERNAL ACTORS

■ WHERE FINANCING AND POLICY ADVICE are needed to make farming environmentally sustainable, traditional international assistance agencies will have important roles to play. For bilateral donors such as the U.S. Agency for International Development (USAID), one of the most critical roles will be to revive support for international agricultural research. For international financial institutions such as the World Bank, money and technical transfers to sustain policy reforms and to support infrastructural investments are also critical. These public sector assistance functions have not been performed adequately of late, and consequently, too large a burden has fallen to nongovernmental organizations (NGOs). Until bilateral donors and multilateral institutions do a better job, NGOs will remain overburdened, and national governments in Asia, Africa, and Latin America will have a continued excuse for their own shortcomings.

Until recently, USAID played a leading role in supporting the 18 international agricultural research centers of the CGIAR (Consultive Group on International Agricultural Research) system. CGIAR research has been a wise investment from an environmental standpoint; it has helped develop the highly responsive seed varieties that allow farmers to raise production without plowing up more land. More recently the CGIAR has moved toward research that benefits the rural environment directly, with biological pest control programs, agroforestry, and breeding for host-plant pest resistance. Despite the success of the CGIAR, USAID has recently cut its core contributions from $48 million in FY1992 to just $28 million in FY1994. Largely because of these weakened U.S. contributions, core funding for the CGIAR system as a whole has fallen by 20 percent. Rural environmental protection and future food production are at greater risk because of this abdication of U.S. foreign assistance leadership.

World Bank lending for all agricultural and rural development projects has also been declining in real terms for more than a dozen years and is now roughly 20 percent lower than it was in 1978, even though populations, food needs, and environmental protection needs—especially in Africa—are growing larger every year.

One reason for these World Bank cutbacks has been a perceived risk that lending for agricultural modernization might attract more criticism from powerful environmentalists, who tend to be suspicious of input-intensive farming in poor countries. Environmentalists have had reason to criticize many World Bank projects in the past, but since 1987 the World Bank has modified a number of its institutions and procedures to take environmental considerations more carefully into account. It will be a shame if these healthy changes go unrecognized by the critics, and if World Bank lending for environmentally critical agricultural and rural development projects continues to falter.

Where governmental and international assistance institutions have failed, NGOs have tried to move in. Local and international NGOs have a proven record for implementing successful grass-roots projects in rural environmental protection. Still, problems of sustainable food production are never going to be solved one project at a time. The states of East Asia did not develop their economies through grass-roots projects implemented by NGOs; rather they implemented land reforms, investments in primary education and health, family planning policies, and economic policies that encouraged private savings and investment, in rural as well as urban areas. One role for NGOs is to speed the emergence of comparably advantageous policies today in Africa, Latin America, and the rest of Asia. Advocacy and public education are also a promising tactic for NGOs within industrial countries, such as the United States. Too many U.S. political leaders carry a vision of environmentally sustainable agriculture that is insensitive to looming food production requirements, clouded with industrial-country biases, and not sufficiently region-specific.

RESOURCE ABUSE AS POWER ABUSE

■ IN THE POLITICAL GEOGRAPHY of sustainable agriculture provided here, at least one deeper cross-regional generalization does emerge. Resource abuse usually reflects an imbalance in power relations, between farmers, between farmers and nonfarmers, or between farm communities and central governments.

Resource abuse by poor farmers using traditional technologies in the wrong place or at the wrong time (first-generation abuse) is frequently an outgrowth of the political weakness of those farmers in relation to other farmers or to their own government. Farming communities with sufficient power to control their own resources are seldom prone to destructive practices. Dangers arise when privileged rural elites and non-accountable governments either take away or threaten to take away local control. Incentives to conserve resources then disappear. Much of the resulting damage (soil nutrient depletion or erosion, deforestation, and depletion of fuelwood supplies) is confined to the farm itself, a further reflection of the political weakness of the farmers in question.

An imbalance of power also helps to explain the second-generation environmental damage done by better off commercial farmers, such as those found in Latin America and East Asia. These well-organized and politically powerful farmers (much like farmers in Europe, Japan, or the United States) are influential enough to demand trade protection and subsidies, which induce an excessive and distorted use of inputs. Much of the resulting damage is felt off-farm (groundwater depletion, downstream chemical pollution), where these powerful farmers escape accountability by wielding influence over local legal and political systems.

Democratization is one answer to these problems of poorly distributed or nonaccountable political and social power in poor countries, but it is not an easy answer. Too often, democratization is initiated and dominated by an urban middle class unconcerned with extending benefits to rural areas. Political and social change must extend to the countryside itself.

Part I
Introduction

COUNTRYSIDES AND FOOD SUPPLIES AT RISK

■ THROUGHOUT THE DEVELOPING WORLD TODAY, agricultural practices are harming the natural environment. Lands not suited to sedentary farming are being cleared and plowed, habitat is being destroyed, biodiversity is being lost, soil nutrients are being exhausted, trees are being cut and burned, watersheds are being degraded, rivers and reservoirs are silting up, irrigated land is becoming waterlogged or saline, surface water and groundwater are becoming polluted with chemicals, and farmworkers are suffering from increased exposure to toxic pesticides, even as pests are becoming more resistant to them.

Farming threatens the natural environment in rich as well as poor countries,[1] but the human stakes are particularly high in the developing world, where the need for food is so acute and growing so rapidly. Roughly 700 million people in developing countries do not have sufficient food to meet their needs for a healthy and productive life. Population growth now implies that the developing world must feed 88 million *additional* people every year, the equivalent of feeding a new Mexico every year or a new India every 10 years. The Food and Agriculture Organization (FAO) of the United Nations has projected that total consumption of cereals for 93 developing countries will have to increase roughly 60 percent between the late 1980s and the year 2010. The necessary production, when considered in cumulative terms, is without precedent. Farmers in poor countries will have to produce more food over the first two decades of the next century than they have produced over the entire 10,000 years since sedentary farming first began.[2]

How can this production be achieved if environmental degradation continues? In much of Africa, where crop yields will have to increase, the "mining" of soil nutrients is now helping to push some average yields into decline. In the Sahel, yields of sorghum have declined about 1.5 percent a year, despite investments in yield-increasing technology. Out of 41 African countries, 13 have registered declines in yield for cereals, and 15 have seen declines in yield for roots and tubers. In much of South Asia today, old irrigated lands are becoming saline and waterlogged and are going out of use almost as fast as new irrigated lands are coming into production. By 1988, the productivity of nearly 20 million hectares of India's cropland

(30 percent of irrigated cropland) had been seriously affected by waterlogging and salinity, forcing farmers to abandon 10 million hectares of productive land. In China and Pakistan, perhaps 15–20 percent of irrigated land is suffering from waterlogging and salinity. From Java to Honduras, trees are being cut, and soils are washing away on newly exposed sloping lands. In western Java, land is being cleared on hillsides that slope more than 25 degrees, with subsequent accumulated erosion as high as 400 tons per hectare. In Honduras, ever larger numbers of destitute subsistence farmers are moving into forest margins and up steep hillsides. As a consequence, soils are being lost at a rate of 10,000 hectares a year, and forest resources are at risk of being totally exhausted, perhaps within a generation. In East Asia, South Asia, and Central America, the natural biological controls for crop pests are being poisoned, even as the pests themselves are becoming more resistant to those poisons. In the Shandong Province in China, deltamethrin was first introduced in 1982 to protect cotton from pests, but within three years pests had become so resistant to it that farmers had to spray their crops as often as twice a month rather than just five or six times a year. In irrigated rice production in Bangladesh, Indonesia, Malaysia, and Thailand, insecticides have actually worsened pest damage by killing too many of the natural enemies of pests.[3]

Given the human and environmental interests at stake, the response of the international policy community to this looming crisis has been painfully weak. Rather than invest more in farming, the protection of farm resources, or agricultural research to innovate new solutions to the problem, the international policy community has recently turned its back on agriculture in developing countries:

- External assistance to agriculture in low-income developing countries declined in real terms between 1980 and 1990—down from $12 billion to $10 billion in 1985 dollars—and agriculture's share of total development assistance fell from 20 to 14 percent over the same period.[4]
- Since 1980, the share of World Bank lending devoted to agricultural and rural development projects has fallen by almost half, and the real value of World Bank lending for projects in agriculture has fallen roughly 20 percent.
- Investments in agricultural research are also shrinking. Core funding for the international research centers of CGIAR (Consultative Group

on International Agricultural Research) peaked in 1990. Since that time, contributions to the centers have fallen roughly 22 percent in real dollars, to a projected $215 million in 1994, even though five new centers have been added to the system; the 13 centers in the system before 1992 have seen their core funding cut 31 percent in real terms.

- Following the lead of the international community, developing-country governments are also spending less money on agriculture. Central government expenditures for agriculture in the developing world have recently averaged only 6–7 percent of total government spending. In roughly half of these countries (including large countries such as Bangladesh, Indonesia, Mexico, and Pakistan), the share of government spending going to agriculture was smaller in the 1980s than in the 1970s.

Instead of doing more, national governments and the international community have been doing less to meet the agricultural crisis in food production and resource protection. Why has their response been so weak?

One significant factor has been a sharp disagreement between agriculturalists and environmentalists over the exact nature of the problem and thus over the proper technical solution to it. Most specialists in agricultural production believe that modern and highly productive farming practices can be made compatible with environmental protection. These agriculturalists associate environmental damage from farming with failure to modernize and hence with the extension of traditional farming practices onto fragile lands or into forests. They usually want farmers to use more purchased inputs and thus produce more food on the lands they already farm.

Many environmentalists, in contrast, see the switch to high-yield farming as part—perhaps even the heart—of the problem, rather than the starting point for a solution. They fear that spreading science-based, high-input farming in poor countries will only lead to more chemical pollution, faster depletion of water supplies, and a dangerous loss of biodiversity. These environmentalists support holding onto traditional or locally evolved farming systems less dependent on purchased inputs or exotic technologies, systems that are more likely to be within the grasp of ordinary farmers, and with proven historical sustainability.

This lack of common ground between agriculturalists and environmentalists deadlocks the policy debate and threatens to paralyze the international assistance community.[5] International assistance organizations, traditionally influenced by agriculturalists but now more fearful of antagonizing environmentalists, have become wary of sponsoring agricultural modernization projects. Yet the number of people needing food in the developing world grows larger every year, and the quality of farm resources continues to degrade.

The only way to break this deadlocked debate is to pay more attention to geography—and to politics. In some regions of the developing world, agriculturalists are clearly right to argue for more use of inputs. In most of Africa, for instance, input use on farms is much too low. Africa's rural environment will not be protected until farmers find ways to substitute higher yields (achieved in part by using irrigation water and inorganic fertilizer) for the cutting of still more trees or plowing still more land. In other regions, environmental advocates are clearly right to argue for less use of inputs. In many Asian countries, the switch to high-yield farming is now so well advanced that a second generation of environmental problems has arisen, mostly linked to excessive or mismanaged use of inputs. Particularly in the industrializing economies of East Asia, prescriptions for reduced use of inputs therefore make sense. In some cases (and especially in some regions), neither camp has the full answer, since movement toward environmentally sustainable farming requires political as well as technical change. In Latin America, inequitable patterns of landownership and rural social relations frustrate the technical advice of both agriculturalists and environmentalists. Here, the rural environment will be difficult to protect unless technical change is accompanied by more fundamental political and social change.

If the debate about sustainable farming can be recast in this region-specific form, and if political relations can be considered alongside technical needs, then agricultural and environmental specialists should find more room to agree. Working together, rather than at cross-purposes, perhaps they will re-animate the policy community. Examining the problem of sustainable farming in a region-specific fashion is analytically demanding but nonetheless essential. A first step is to stop making any reference at all to the developing countries, the South, or the Third World. These carelessly aggregated labels need to be set aside, if safe technical and prescriptive generalizations are to be found. Accordingly, this essay makes no further reference to "the developing countries." Instead, it offers technical prescriptions on a region-by-region basis or within even more narrow geographic and agroclimatic confines.

First, the essay presents a three-part region-by-region diagnostic and prescriptive review examining Asia, Africa, and Latin America in turn. Each section ends with a summary of remedial actions waiting to be taken by governments within the region. These regional reviews are followed by a discussion of remedial actions waiting to be taken by the international assistance community, including the U.S. Agency for International Development (USAID), the World Bank, and nongovernmental organizations (NGOs). The essay concludes with a review of at least one deeper (albeit nontechnical) pattern common to all geographic regions: a tendency for the abuse of agricultural resources to grow out of excessive concentrations of political power.

Part II
Asia

FIRST- AND SECOND-GENERATION ENVIRONMENTAL DAMAGE FROM FARMING

■ Rapid, equitable, and sustained economic growth sets much of Asia (especially East Asia) apart from both Africa and Latin America. In Asia, strong gains in food production, made possible by science-based, input-intensive, green revolution farming technologies, have contributed significantly to this success. China's total production of grain has almost tripled since 1960. Indonesia's production of rice nearly doubled during the 1970s and the 1980s alone. In India, total production of food grains is nearly three and a half times higher than it was in 1952.

Agriculturalists cheer this success because they claim that it has fed people at the lowest possible cost to the rural environment. Clearly, it has helped to contain a first generation of rural environmental problems associated with the practice of extending traditional farming practices onto new and more fragile lands. Yet it did so at the cost of creating a second generation of environmental problems primarily associated with overuse or mismanagement of farming inputs, such as irrigation water and chemicals.

Environmentalists argue that Asia's escape from first-generation rural environmental problems is not sustainable and, hence, is not a model for others to follow. This study will argue the need to respond to second-generation problems, but not through a return to traditional farming technologies. Moreover, responses must account for variations within the region. Only some parts of Asia enjoy the luxury of worrying about the second-generation environmental problems that accompany agricultural success; many less-favored dryland or upland regions have yet to achieve a green revolution and are therefore still struggling with first-generation environmental problems plus the even more urgent problem of lagging food production.

FIRST-GENERATION PROBLEMS

Asia is the original home both of sedentary agriculture (invented in South Asia roughly 10,000 years ago) and of science-based, input-intensive green revolution agriculture. China had something of an early green revolution in the eleventh century, during the Sung Dynasty, when it introduced new varieties of quick-ripening rice into the irrigated Yangste Delta.[6] The successful operation of intensive farming in Asia over the ten centuries since that time seems to offer strong evidence that it is environmentally sustainable.

Asia's past, however, is an imperfect guide to its immediate future, given this century's astonishing revolution in human population growth. For most of its history, Asia's population was relatively dense but also relatively stable; rapid growth was contained by death from widespread disease plus periodic famines brought on by war or drought. All this changed in the middle of the twentieth century, with the introduction of modern medical inoculations against childhood diseases and improved public health and sanitation practices. The awesome demographic implications were at first underestimated, even by specialists. At an international population conference held in Paris in 1937, estimates were presented of the population of Java by the end of the century; these estimates were surpassed in only 20 years, not 60.

Asia's mid-century surge in population put the environmental sustainability of traditional farming systems under unprecedented stress. Total consumption of grains began doubling every 25 years, and (unlike either Latin America or Africa) little additional land was available to bring under crop production. Some lands cultivated in the past were even abandoned by farmers due to degradation. Under these circumstances, the only way to meet new production requirements, without plowing up still more fragile dryland or without cutting into easily degraded upper watersheds and forest margins, was to boost yields even higher on good irrigated lands already under crops.

East Asia (roughly from Thailand eastward), with a long history of successful farm intensification based on innovative plant breeding and heavy use of inputs (including human labor as well as water and nutrients), responded successfully to the challenge of rapid population growth. Particularly during the decades of the 1960s and 1970s, massive investments were made in crop research and irrigation infrastructure, especially for the production of basic food grains such as rice and wheat. As a consequence, Asia's annual growth of rice output increased from 2.1 percent in 1955–1965 to 2.9 percent in 1965–1985. Between 1965 and 1985, paddy rice yields (tons per hectare) in Taiwan and Korea increased 53 percent and 132 percent, respectively. The broadly based gains in nutrition, employment, and social welfare that accompanied Asia's late twentieth-century green revolution have been carefully documented.[7]

The rural environmental gains from this green revolution have also been dramatic, yet they are much harder to measure, since they take the form of damage that did not occur. Higher crop yields helped Asian farmers avoid a further expansion of crops into forests or onto fragile and sloping lands. Area expansion contributed nearly one-third of Asia's growth in rice output as late as the 1960s, but thanks to gains in yield, it became practically unnecessary by the 1980s.[8] In Indonesia specifically, for every increase in yield of 100 kilograms per hectare in wet paddy rice, about 4.6 percent of the country's badly endangered forest cover was saved from shifting cultivation.

Parallel environmental gains were achieved in South Asia, especially by breakthroughs in wheat yields. In India, in the rich and well-watered wheat-growing regions of Punjab, high fertilizer use in combination with newly available seed varieties made possible a rapid growth in yield and a doubling of total production. For India as a whole, by 1970–71, wheat production was almost twice as high as it had been in 1964–65, total stocks of food grains grew to more than 8 million tons, and India was able to terminate its large and politically troublesome imports of food aid from the United States. Rice production eventually increased as well, more than doubling in Punjab and Haryana between 1971 and 1976.

India's successful adoption of high-yield farming averted a widely predicted famine. It also saved the country from a rural environmental calamity that would have been experienced on the way to that famine.[9] If India had relied on traditional low-yield farming techniques to increase total production in the 1960s and 1970s, it would have had no choice but to cut more trees, destroy more wildlife habitat, and plow up more fragile sloping and dryland soils. In 1964, India produced 12 million tons of wheat on 14 million hectares of land. In 1993, thanks to the green revolution, India produced 57 million tons of wheat on 24 million hectares of land. To produce this much wheat at the 1964 level of yield, India would have had to plant roughly 60 million hectares. In other words, the green revolution allowed India to meet its evolving food needs *without* plowing an additional

36 million hectares of cropland for wheat. "Thanks to plant breeding," concludes M. S. Swaminathan, at India's Centre for Research on Sustainable Agricultural and Rural Development, "a tremendous onslaught on fragile lands and forest margins has been avoided."[10]

Few environmental critics of India's green revolution address the issue of how India might otherwise have produced enough to feed its rapidly growing population. Scientist and activist Vandana Shiva's book-length polemic against the green revolution in India makes no reference to the growth of population or consumption.[11] Although critics should worry about the second-generation environmental problems that green revolution farming tends to create, they should not ignore the large first-generation problems—both environmental and nutritional—that high-yield farming has so far helped to solve.

Asia's environmental gains from high-yield farming are nonetheless attended by at least three serious limitations. First, the gains in production are not shared evenly by all farming regions in Asia, or by all farmers. Second, although these gains protect land, trees, and habitat from further encroachment, they often bring a second generation of rural environmental problems, such as chemical pollution and loss of biodiversity. Third, these gains can now be sustained only through continued public sector investments in research and infrastructure, which governments in the region—and also international donors from beyond the region—seem reluctant to provide. Each of these limitations is considered in turn.

EXCLUDED BENEFICIARIES. Not all farmers and farmlands in Asia participate in the region's successful strategy of agricultural intensification. Where the strategy is not applied, or where it is inequitably applied, rural environmental problems of the first-generation kind often continue or are compounded. Barriers to extending this strategy are encountered in countries with large populations of landless rural laborers, where farmers have inadequate and insecure land tenure; in less-favored regions of mostly upland or dryland farms; and also in countries where governments are not accountable to local farm communities.

In much of Asia, especially South Asia, many landless rural dwellers earn income through environmentally destructive actions such as cutting and selling fuelwood from previously protected common property areas. Bangladesh, India, and Pakistan together have more than 30 million landless rural households, representing close to 40 percent of total rural households. Because farm size is already so small in most of Asia, the problem of rural landlessness will worsen as the population continues to grow.

Even when farmers have access to land, if that access is not legally or socially secure, they may employ practices that damage the environment. Farmers may not wish to make long-term improvements on lands owned by somebody else. Where landownership is required (for collateral) to borrow money, they may lack the financial means to make such improvements or even to intensify production. China, the Republic of Korea, and Taiwan minimized such problems by completing ambitious post-World War II land reforms, but in many other Asian economies, inadequate or insecure access to land remains a severe problem.

In India's Punjab state, where most land is cultivated under the direct personal supervision of owner-farmers, investments in intensification came quickly in the 1960s; yet in the state of Bihar, at the eastern end of the same fertile Indo-Gangetic river valley, intensification lagged, partly due to the prevalence of semi-feudal relations of production, especially sharecropping. Sharecroppers who must give the landlord half or more of what they produce have little incentive to spend their own money to purchase fertilizer or to improve the land with irrigation. And the absentee landlord, who may come from an urban-dwelling upper caste family that has traditionally held land for social status or as a hedge against inflation, may be equally slow to undertake productivity-enhancing investments. Partly as a consequence, farmers in Bihar apply too little fertilizer, use too little groundwater, and produce lower yields than farmers in Punjab.[12]

Farmers with an insecure hold on the land also tend to manage low-input farming systems improperly. In Thailand, farmers without secure title cut trees in upland areas, exhaust the soils, and then move still farther into the forest before their land can be taken away. In the Philippines, already about 18 million people (one-third of the total population) live in upland areas, and roughly half of them now farm on fragile forestlands without secure tenure and hence with little incentive to protect resources. They cut trees, expose erodible soils, and degrade watersheds—practices that cause floods, water shortages, and degradation of good irrigated farmlands downstream. Degradation of the upper watershed leads to the loss, downstream, of roughly 7,000 hectares of irrigated land every year, with an implied annual incremental loss of production of 24,000 tons of rice.

On the island of Java, population density is now about 800 people per square kilometer, as high as in Bangladesh but in a setting where the terrain allows less land to be cultivated. Average farm size is down to less than 0.5 hectare, and nearly 40 percent of the population dependent on agriculture is effectively landless. Attempting to cope with this crisis, the Indonesian government, since 1980, has sponsored resettlement programs that aim to move as many as 10 million families to the nation's relatively underpopulated outer islands by 2000. One unfortunate result is that resource-destroying farming practices are introduced into the more fragile regions of these other islands. The transmigrants cut trees with chain saws, plant upland rice on erodible soils not suited to such crops, and attempt to force irrigated rice systems into areas prone to rapid deterioration.

PHYSICAL LIMITATIONS. Also blocking the extension of a yield-boosting strategy is the difficulty of moving irrigated farming into Asia's vast upland and dryland areas. Farming on nonirrigated rainfed lands is still the principal source of livelihood for roughly 65 percent of the region's rural population, including most of the poor. Asia's upland watershed areas, especially those in the Hindu-Kush-Himalaya region, stretching from Pakistan across northern India, Nepal, Bhutan, and China, are generally too hilly and too remote for irrigated farming and are prone to erosion from rainfed farming and earthquakes. Historically, these upper watershed areas were under forest cover, but over time, and under population pressure, they have been exploited for timber, fuel, fodder, and various destructive forms of agriculture. Now only 19 percent of the area remains under forest.

Rather than intensify production with more labor or purchased inputs, farmers continue expanding the area (and hence the area of destruction) under traditional cropping practices. In the hills of Nepal, only 6 percent of the rice crop consists of high-yield varieties. Much of the environmental damage from this pattern of upland farming is felt downstream, where eroded materials are deposited in the reservoirs and irrigation systems that are critical to lowland farming. In India, siltation reduces the life of many reservoirs by more than 75 percent. Excessive surface runoff, flash flooding, and reduced stream flows during dry seasons threaten downstream farm and nonfarm populations alike.

Parts of Asia, especially in the southwest, also contain extensive areas of lowland farming that receive too little rainfall and are under increasing threats of damage from overgrazing or mismanaged irrigation systems. In these regions, river systems are weak and scattered, underground aquifers are either too deep or recharge too slowly, and adverse soils often combine with seasonal extremes of drought and flood. The uncertainty of rainfall makes it difficult to persuade farmers to invest in improved water management, since the improvements may be unnecessary in wet years, and in very dry years nothing works. History shows that imprudent use of such drylands can do lasting environmental damage. In ancient Mesopotamia, the diversion of river water onto dry valleys led to severe problems with silt and salt, eventually bringing down the Sumerian civilization.

Asian populations on these drylands are surprisingly dense: three or four times as dense as populations in the semi-arid regions of West Africa. Most of Asia's poorest people now live on these nonirrigated lowlands in India, Bangladesh, and the Terai region of Nepal. Most of the cheap or easy opportunities to irrigate such lands have already been taken, and further irrigation is often prohibitively expensive, environmentally dangerous, or both. In Southwest Asia, 70 percent of all potentially irrigable land is already irrigated.

In some of the more densely populated river basins of arid and semi-arid Southwest Asia, abundant surface water and rich alluvial soil should make canal irrigation an attractive option. Yet so much environmental damage has accumulated from the mismanagement of *past* irrigation schemes that the challenge is not so much to extend irrigation onto new lands as to prevent further degradation and abandonment of land already irrigated. In rural Pakistan, 25 percent of irrigated land is saline due to high water tables, and 60 percent of this is already unsuitable for cultivation.[13] In India, because of environmental degradation plus rapid increases in recurrent costs for operation and maintenance partly due to excessive centralization, the sustainability of irrigation systems is in question on financial grounds alone. Without greater transparency, accountability, and institutional decentralization—for example, through water user associations controlled by farmers—neither food production nor resource protection can be assured.

POLITICAL SHORTCOMINGS. Lack of accountability of governments to local farming communities is a third barrier to extending rural environmental gains. Especially in South and Southeast Asia, national governments too often exacerbate the difficulty of protecting rural environments by disrupting local systems of common property, abusing publicly owned forestlands, and locating large environmentally damaging projects, such as hydroelectric or irrigation dams, in fragile but populated rural areas.

The local systems of common property that traditionally protected important environmental resources, including pastures, forests, rivers, and village ponds, are under intense pressure in Asia. One reason is the rapid growth of the rural population, which tends to shrink individual shares of a protected resource area, thus reducing individual incentives to protect resources cooperatively (especially in the face of outside poaching, which increases with population density). Too often, however, governments that are centralized and not accountable to local communities exacerbate the bad situation. Rather than seeking to protect common property systems, central political authorities too often privatize those resources inequitably or absorb them into the public sector.

The demise of well-functioning common property regimes in India is a case in point. In a study of 82 villages across seven states in dryland India, the total area being managed as common property declined 30–50 percent between the early 1950s and early 1980s. During the same period, effective management of the remaining common property areas weakened (only 10 percent of villages still regulated grazing or provided watchmen, none levied grazing taxes, and only 16 percent still obliged users to undertake maintenance and repairs), and as a consequence physical degradation became widespread.[14] One reason was that the Indian government took an aggressive role in privatizing common property resources in the early

1950s. This was done in the name of helping the poor, but poor households received, on average, only about half to one-third the amount of land given to more prosperous households, and most poor households soon were dispossessed of one-quarter to half of the land allocated, often because it was too little (or its quality was too low) to provide them with a workable farm. In other instances, the state used its power to monopolize the collection or marketing of products, such as trees, from common property areas. Villagers protested and even attempted to defend their traditional rights in court, but to little avail.

Often just the threat that central authorities might disrupt locally controlled resource management schemes is enough to break up those schemes, engendering an "exploit it now before they take it away" attitude. Unless new institutions can restore a secure sense of local control (such as pastoralist associations, water user associations, or tree management collectives), former common property areas will continue to be exploited by central authorities or will degrade into open-access areas that engender uncontrolled abuse of resources.

The detrimental role of non-accountable central government is seen most clearly in the abusive attitude taken in many parts of Asia toward publicly owned forestlands. The loss of forest in Africa and Latin America can frequently be traced to private activities, including shifting cultivation and cattle ranching (sometimes induced by government subsidies). The loss of forest in Asia, however, can more often be traced to direct actions by governmental authorities: colonization and area development programs (such as the transmigration program in Indonesia), officially sanctioned expansion of cash-cropping (cassava for industrial purposes in Thailand), and most of all officially sanctioned commercial logging. The once-dominant commercial log exporters, Philippines and Thailand, have exhausted their forests in this fashion, and the struggle to halt commercial deforestation in Asia is now being waged in the few countries with exploitable forest lands remaining: Indonesia, Laos, Malaysia, and Myanmar.

Commercial forests in Asia are almost invariably under state ownership or control, frequently a legacy of the colonial era. In Indonesia, the Forestry Department has jurisdiction over 144 million hectares, or roughly 75 percent of the country's entire surface area. Asia's highly centralized forest agencies work closely with forest resource industries;

they usually operate from a short-term perspective and are seldom accountable to the forest-dwelling people who often hold traditional rights to forest resources under customary law. An estimated 200 million indigenous and tribal people live within and depend on forests in South and Southeast Asia, but in many cases their presence is not even acknowledged by the national bureaucracy that has assumed legal jurisdiction over the land.

Forestry departments in Asia seldom manage public resources in a publicly accountable fashion or with more than short-term foreign exchange earnings in mind. They may be pressured from above to enter into corrupt arrangements with local concessionaires who have close personal or political ties to the nation's top leadership. International companies can also promote mismanagement, but they are not always the first source of irresponsible behavior. In Indonesia, local concessionaires are sometimes reluctant to permit their joint-venture multinational partners to make long-term investments in reforestation, and government ministries frequently encourage this short-term attitude.[15] By one estimate, the short-term social cost of the government's underpricing of forest wood in Indonesia is approximately $500 million annually. The commercial and environmental losses (especially loss of biodiversity) are obvious, but sustainable agriculture is also sometimes the victim. On Java, off-site soil erosion from degraded upland forests, as well as from rainfed agricultural land, costs farmers $75 million a year.

Overly centralized and nonaccountable governments also damage the rural environment through their penchant for constructing large rural infrastructural projects, especially big dams. Of all the large dam projects (dams over 15 meters in height) financed by the World Bank since 1970, almost 40 percent are in South Asia alone, and another 17 percent are in East Asia. The environmental disruptions that accompany such projects are notorious. Some affect farming, including the inundation of productive farmland and the resettlement of displaced populations onto inferior or more fragile lands, yet many are largely external to agriculture, such as disruption of fisheries, impacts on wildlife, or induced seismic activity. Irrigation for farming is often a secondary purpose in the construction of big dams; electric power, water supply, and flood control for urban areas are more likely to be the favored political objectives. The controversial

Three Gorges Project in China is designed for hydropower, flood control, and navigation on the Yangtze River, not for irrigation.

Most environmental critics of big dams focus on the past role of the World Bank in helping to finance construction. Currently, however, the World Bank is a major funding agency for less than 5 percent of the average annual construction of dams worldwide, and today's single most controversial dam project in South Asia—the Narmada Valley Development Program in India—is no longer funded by the World Bank. The political motive behind most rural mega-projects now comes from central government authorities, rather than foreign assistance donors or advisors.

In some cases, grass-roots campaigns are pressuring governments to modify or halt environmentally damaging dam projects. In Thailand, the 1982 Nam Choan Project (a hydropower project, which the government had proposed to locate in a wildlife sanctuary) was eventually halted because of public pressure. Public protests of this kind are most likely to be effective if they are mobilized early enough in the proposal stage, before the government has fully and publicly committed itself to the project (this point has now been passed in the case of Three Gorges and Narmada). In countries with authoritarian governments, this "public protest" path to political accountability is a weak foundation for protecting the rural environment, but it is gaining strength, often with assistance and intervention from international environmental NGOs.

SECOND-GENERATION PROBLEMS

The first-generation environmental problems noted so far are most severe in the parts of Asia where green-revolution farming has not yet spread. Where yields have been successfully increased, especially in the wet lowland rice bowls of East and Southeast Asia and in the irrigated grain belts of South Asia, an entirely different set of second-generation environmental problems is rapidly emerging: excessive use of water, inadequate containment of nutrients and waste, pesticide poisonings and pest resistance to chemicals, and loss of biodiversity.

Irrigated high-yield farming saves the need to plow fragile land, but if water use is mismanaged (often because of unwise government subsidies), this kind of farming can eventually become environmentally unsustainable even on good land. In Punjab in northwest India, the government subsidizes 86 percent of the electric bill for pumping irrigation water as a reward to politically powerful commercial farm interests. As a consequence, farmers pump so much water that groundwater tables are dropping at an unsustainable rate of about 0.8 meter a year. In the semiarid regions of southern India, where subsidized pumping of groundwater allows production of three rice crops a year, water levels are now dropping nearly a meter every year.[16]

Excessive use and poor containment of nutrients is another environmental problem on high-yield Asian farms. Compared with the United States, China uses 2.7 times as much fertilizer per hectare of cropland, Korea uses 4.2 times as much, and Taiwan uses 5.1 times as much. When purchased fertilizer becomes more affordable in the rapidly growing countries of Southeast Asia, they too may go down this nutrient-intensive path; between 1970 and 1990, Thailand's fertilizer use per hectare increased nearly sevenfold. Subsidies are frequently to blame for excessive use of nutrients. When Indonesia subsidized fertilizer purchases by as much as 68 percent, between 1980 and 1985, consumption increased 77 percent. So long as nutrients can be contained in the farmer's field and taken up in the roots of plants, they do little environmental damage. If they leave the farm field as surface runoff, however, or if they leach into groundwater, the off-farm environmental damage can be serious. Nitrates in drinking water can place human health at risk and cause eutrophication of streams and ponds.

Excessive use of chemical pesticides, particularly by producers of irrigated rice, wheat, and cotton, is an even more immediate health and environmental problem. Farmworkers who spray these poisons in the fields, often without being adequately informed about or protected from the dangers, are the first victims. In Central Luzon, in the Philippines, data over a 20-year period (1963–1983) and circumstantial evidence (high death rates during peak spraying times) strongly suggest that a 27 percent increase in mortality can be directly attributed to insecticides.[17] Destruction of wildlife, including commercially valuable fish, amphibious, and terrestrial species, is also severe. In Java, the traditional practice of *mina-padi* (producing fish within rice) has almost disappeared due to water pollution.

The heavy use of chemical pesticides on irrigated rice is not only dangerous and costly, but also frequently unnecessary and counterproductive. The 800 or so insect species that live in a typical irrigated rice field can usually, if given a chance, protect crops against plant pests better than any heavy application of industrial pesticides.

Indonesia provides a particularly good example of the counterproductive use of pesticides. That country began to apply chemical pesticides heavily in the 1970s, badly upsetting the natural balance. Brown planthoppers, which feed on rice, began to multiply in large numbers when their natural controls (such as spiders) were killed by insecticide. Not realizing that pesticides had largely created the planthopper threat to rice, the Indonesian government offered farmers an 85 percent subsidy to spray even more, which exacerbated the problem, as the planthoppers began developing stronger resistance to the chemicals. Finally, in November 1986, a presidential decree banned the use of 57 broad-spectrum insecticides for rice and declared integrated pest management to be the preferred strategy for controlling pests. Under integrated pest management, farmers rely heavily on the natural enemies of plant pests, carefully monitor pest populations, and use chemical poisons only when pest populations exceed a relatively high economic threshold. Indonesia's switch to integrated pest management was an instant success. The use of insecticides fell 90 percent (saving farmers money), and natural predator populations promptly recovered, which allowed average rice yields to increase. By eliminating its pesticide subsidies, the Indonesian government also saved more than $100 million a year in budget outlays.[18]

Unfortunately, integrated pest management in Asia has not spread widely beyond Indonesia. As of 1990, only about 4 percent of Asia's farmers of irrigated rice were practicing the strategy, and outbreaks of brown planthoppers have recently followed heavily subsidized applications of insecticides on rice in Bangladesh, Malaysia, and Thailand. Governments subsidize chemicals out of habit, to maintain well-established patronage relations between ministries and farmers, and to support national (often state-owned) chemical companies. Since 1977, the FAO has made significant efforts to develop and spread the technologies of integrated pest management throughout Africa, Asia, and Latin America, but donors have offered only narrowly based and inadequate support for these efforts, and agricultural ministries at first showed little interest. In response to a 1986 FAO questionnaire, 70 percent of all governments that responded said they were not promoting integrated pest management.[19] Although national commitments have more recently improved, international efforts to create a new global Integrated Pest Management Facility in 1994 (to harmonize diverse efforts to promote integrated pest management within FAO, the United Nations Development Programme, the United Nations Environment Programme, or UNEP, and the World Bank) correctly stressed the continuing urgency of having a supportive enabling policy environment in each country.

A related problem with high-yield production of food grains is the loss of biological diversity in farming. Many of the improved varieties of rice used in Asia today are biological descendants of IR-8, the first "miracle" variety released by the International Rice Research Institute (IRRI) in 1966. When rice varieties with a narrow genetic base are grown in monoculture over vast areas, the risk of a rapidly spreading disease or infestation is considerable.[20] Until now, these risks have been contained through the continuous efforts of plant breeders (many still working at IRRI in the Philippines) to offer a steady stream of new varieties, with changing resistances and tolerances. New varieties tend to replace old varieties in Asia's rice fields every five to ten years. If this plant breeding treadmill were to break down, Asia's highly intensive rice farming systems would soon be placed at considerable biological risk.

CAN GROWTH OF YIELD IN IRRIGATED RICE FARMING BE SUSTAINED?

■ SOME SCIENTISTS ARGUE that Asia's plant breeding treadmill is already beginning to falter, under the increasing pressure that high-yield farming places on that region's base of resources. The data are not yet conclusive, but some research scientists now warn that the recent growth in yield achieved in Asia's irrigated rice fields is not likely to continue. This is primarily because physical yield potentials, established on research test plots with high use of inputs and ideal water management, have not been substantially increased since the 1960s. There have been research breakthroughs in yield stability and pest resistance, but not in genetic yield potential.[21]

Actual yields on Asian farms now average only about half of current yield potential, seeming to imply considerable "catch up" options; however, the best lands (roughly 90 percent of the irrigated area) already use high-yield varieties, so the easiest part of the catch up process has already been completed. Expanding irrigation will also be increasingly difficult and expensive, because the easiest sites have already been irrigated. Adding another 100 million hectares of irrigation could cost from $400 to $800 billion at today's prices, thus running up against financial as well as physical limits in most countries.[22]

A new concern, moreover, is that the yield potential of heavily cropped and long-irrigated land may itself begin to decline. Whereas traditional producers of irrigated rice grew only one crop of rice a year, followed by a fallow period during the dry season, intensive growers now produce two or three consecutive rice crops a year on the same land, adding purchased inputs of fertilizer as necessary. According to cropping experiments in the Philippines, fields that have been farmed continuously in this intensive fashion since the 1960s, using the best available rice cultivars and the best scientific management, are experiencing declining yields over the long term when input levels are held constant. In some parts of the Pakistani Punjab, stable average yields of modern wheat varieties have required a tripling of average fertilizer doses, plus a continued introduction of new varieties, over the course of two decades.[23] Such problems are almost certainly a result of the gradual degradation of continuously cropped (and sometimes constantly flooded) high-potential farming lands.

What is being done to help Asia's high-yield farmers escape this new constraint on their performance? Since continued population growth and steadily higher consumption needs eliminate the possibility of growing only one rice crop a year, major new investments are clearly needed in rice genetics and rice production research. If current knowledge is not adequate and if old knowledge would be a step backwards, then this should be the time to invest in new knowledge.

Surprisingly, given the importance of this emerging worry, the annual growth in spending for agricultural research in Asia has been declining in real terms (from a 7.4 percent growth rate in the 1960s and 1970s to just 4.6 percent in the 1980s). Agricultural lending and assistance to the region has also been declining. Lending and assistance to Asia as a whole, and also specifically to less wealthy South Asia, peaked in real terms in the late 1970s and early 1980s and by the late 1980s had declined roughly 50 percent. Funding for the IRRI has fallen in real terms more than 20 percent since 1990. The decisions of external donors to reduce their investments in research have been mirrored by those of national governments.

For all of these reasons, it cannot yet be said that Asia has solved its environmentally sustainable food production problems, even on its highest potential lands. And in those parts of densely populated Asia where land and water systems are not suited to high-yield farming, the solution is even more distant.

SUMMARY PRESCRIPTIONS

■ How can the second-generation environmental problems associated with high-yield farming in Asia be reduced or better managed? First, governments need to reform the input subsidy and trade protection policies that distort price signals and encourage overuse of water and chemicals. This will be politically difficult, given the universal tendency of commercial farmers to become better organized and more powerful within the political systems of nations that undergo rapid industrial transformations.[24] In today's most advanced industrial countries in North America and Europe, however, policies such as separating payments to farmers from production outcomes ("decoupling") have begun to break the link between politically necessary income support for farmers and damaging incentives to use more inputs. Industrial countries are also providing more of their income support to farmers in the form of incentives that explicitly protect the land, reduce acreage, and reduce input use. Farmers themselves are reducing their expensive and environmentally damaging use of inputs, especially fossil fuels and pesticides, by employing improved farming techniques, such as the no-till alternative to deep plowing or the better timed and more precise application of smaller quantities of chemicals. Having earlier substituted the use of inputs for the use of additional land, these farmers in the industrial world are now substituting improved management for inputs.

There is some hope that as Asian countries become steadily more wealthy and as farmers gain access to new technical options, they will be able to adjust in this fashion. In Japan, total consumption of pesticides is already in decline, and concentrated hog production facilities are being downsized. Taiwanese hog farmers are being offered low-interest-rate loans to invest in wastewater treatment facilities, and pork producers are required to spend some 20–30 percent of their net returns to clean up animal waste. The Taiwanese government is using financial incentives to sponsor a transition in the countryside to what it calls leisure agriculture and tourist farming.[25] And late in 1993, both Japan and Korea (as a part of the Uruguay Round of multilateral trade talks in the General Agreement on Tariffs and Trade) finally agreed to open their highly protected rice markets to imports, which will reduce the price distortions that encourage the excessive use of inputs in these two countries.

Second, if industrializing Asia is to manage its pending rural environmental crisis successfully, more scarce land and water resources will have to be shifted out of the production of low-value food grains and animal feeds and into the production of higher value commodities such as fruits, vegetables, and animal protein products (which consumers demand more and more as their incomes rise). Without placing a heavier total burden on the environment, such products can generate more income to farmers and more employment to accommodate a still-growing population. Food grains and animal feeds are needed in Asia, but as income rises (and particularly as foreign exchange is earned through the production of exportable manufactured goods), Asia should begin to import a growing share of its food grains and animal feeds from abroad. It makes environmental sense for the industrializing economies of Asia to import more wheat, corn, soybean products, and even rice, since the principal exporters of these products (such as the United States) use many fewer chemicals per acre than East Asia. It also makes commercial sense. Today's most successful East Asian economies—Japan, Taiwan, and Korea—are already the region's largest importers of farm products. In FY1992, the

United States exported $8.3 billion in agricultural goods to Japan and another $4.9 billion in farm goods to the other nations of East Asia.[26] Some now argue that the future food needs of China will be too large for the world's food system to satisfy in similar fashion through trade.[27] Such arguments about the limited capacity of the world's food trading system to provide affordable commercial exports have always been wrong in the past.[28] The main point here is that dangers of food shortage and environmental damage in countries like China will be much larger without trade than with trade.

Third, income and employment off the farm need to continue growing, both in urban and in rural areas. Income must increase to create adequate commercial demand (initially around urban centers) for the higher value crops that Asians will have to grow in order to get adequate income and employment returns from their constrained land and water resources. Higher off-farm rural employment will make an important contribution, because it gives farm families a chance to boost their income while doing less farming plus a financial means to afford resource-protecting investments when they do farm.

At present, 700 million people live in absolute poverty in Asia. Their hopes for the future depend on continued economic growth. Keeping them poor will not protect the environment, at least not the rural environment. Economic growth based on increasingly efficient use of resources, including a reduced and better managed use of many farm inputs, offers greater hope. If done properly, urban industrial and post-industrial development can improve the environment of densely populated Asia. Especially if the region's rapid economic growth is used to finance investments in today's cleaner urban technologies, development of the industrial and service sectors does not have to be heavily polluting; in 1991, Japan produced twice its 1973 output using no more energy and significantly *reducing* total emissions.[29]

For Asia, then, it is an essential first step to distinguish between those parts of the region that have intensified their production and those that have not.

In parts of the region where high-yield farming has already been achieved, use of inputs such as water, fertilizer, and pesticides needs to be cut back, better managed, or better contained. In many cases, this will require moving away from distorting farm support policies such as commodity import restrictions and input subsidies. Where farm income supports are necessary for political reasons (as they are in the industrial world), they should either be "decoupled" from incentives to boost inputs or be conditioned directly on better management practices. More public spending on research (both national and international) is also needed for two ends: 1) to help these countries innovate and extend new farm management techniques, such as integrated pest management, that can substitute for chemical inputs and, most important, 2) to generate another round of gains in actual yield potential as well as in pest and disease resistance for basic food grains such as rice. Tighter command and control regulations are also required, as are incentive policies, inducements, and research investments, to reduce occupational risks from pesticide use and to clean up animal feedlots.

In other parts of the region where high-yield farming has *not* been achieved, containing the environmental costs of the green revolution is not yet as important as containing population growth, while launching as much of a green revolution as biophysical limitations permit. In these areas, much larger investments are needed in family planning, in the spread and maintenance of irrigation, in the improved availability of key inputs such as new seeds and fertilizer, and in the construction of basic rural infrastructure, especially farm-to-market roads. Just as important in these less-favored areas, new governmental attitudes must be developed toward rural farming communities. Where common property systems are still functioning, central bureaucratic authorities should respect their operation. Where such systems are no longer functioning, or where large numbers of rural dwellers are now landless, government policy should provide equitable and secure local access to land and water. Government policies toward land, forests, and watersheds must, above all else, be based on principles of local participation and secure local control.

Part III
Africa

FIRST-GENERATION ENVIRONMENTAL PROBLEMS AS A CONSTRAINT ON FUTURE PRODUCTION OF FOOD

■ WHILE GREEN REVOLUTION FARMING has brought numerous second-generation environmental problems to Asia's rural areas, most of Africa has yet to achieve a green revolution. As a consequence, Africa (that is, Sub-Saharan Africa) is still struggling to solve first-generation problems associated with expanding low-input farming onto fragile lands or deeper into forest margins.

Africa's first-generation environmental problems are now so serious that they threaten to worsen the region's already grave crisis in food production. Total agricultural production increased at an average annual rate of only 2.1 percent during the 1980s, well below the region's 3.2 percent rate of population growth. Africa is the only region of the world in which per capita production of food has been *declining*. These adverse production trends have already resulted in more hunger, offset only in part by higher imports of food aid from other countries. During the 1980s, average daily caloric intake was just 87 percent of requirements. As much as 40 percent of the population of Sub-Saharan Africa does not obtain enough food.[30]

These trends reflect a general failure to boost crop yields. While yields per hectare for basic food crops grew in all other nonindustrial regions during the 1980s, in parts of Africa yields actually *declined*. In the Sahel, yields of sorghum declined about 1.5 percent a year, despite investments in yield-increasing technology. Out of 41 African countries, 13 have registered declines for cereals, and 15 have seen declines for roots and tubers.[31]

Drought contributes to this pattern of yield decline, but resource degradation is a factor as well. FAO has estimated that nitrogen and phosphorus are being mined from African soils (and not replaced) at rates of roughly 44 million and 12 million tons, respectively, every year. Zimbabwe alone loses on average 1.6 million tons of nitrogen and 240,000 tons of phosphorus each year. The base of resources is also being degraded through erosion, salinity, and sedimentation. North of the equator, more than 35 percent of the land is affected by either erosion or salinity. Sediment in many rivers is increasing half again as fast as the populations in their catchment areas. Sediment levels are increasing approximately 5 percent a year in Nigeria, Tanzania, and Zimbabwe; 4.3 percent in the Sudan; and 3.2 percent in Tunisia.

On Africa's drylands, the greatest threat to farm production is a unique form of degradation commonly called desertification.[32] Because of wide fluctuations in rainfall and poor historical data, it is difficult to determine the extent to which this problem is human-induced and irreversible. Still, on the southern edge of the Sahara, some 650,000 square kilometers of once-productive land—an area the size of Somalia—have become desert over the past 50 years, and FAO has estimated that an average of 50,000–70,000 square kilometers of land in Africa go out of production every year. If recent trends in the depletion of ligneous vegetation continue in the Sahel region, most of the "browse" will disappear in the next 20 years, at which point the cattle and people that depend on them will not survive the dry season.

Since about 1970, part of the problem has been a mysterious pattern of below-trend rainfall in the Sahel (possibly associated with nearby patterns of deforestation). In Côte d'Ivoire, which has the region's highest rate of deforestation, average annual rainfall has been in an accelerating decline since the 1970s; in Senegal, annual rainfall has decreased 2.2 percent yearly over the past two decades.[33]

Whatever its role in this decline, deforestation is proceeding at a rate 30 times faster than reforestation, with roughly 5 million hectares lost every year, primarily as a result of clearance for agriculture.[34] In West Africa, roughly 4 percent of the closed forest is razed each year. The fragile soils exposed are then rapidly degraded or lost; rates of soil erosion can increase many fold on steep slopes following loss of tree cover. Tree materials that once protected soils from leaching and erosion, and were once returned to the soil as nutrients, are now being grazed by animals, burned for fuelwood, or used by farmers as building materials. A large quantity of biomass is simply burned in place to clear land, provide green herbage for livestock, or flush out wild game. For the continent as a whole, about 25 percent of the semi-arid zone, 40 percent of the subhumid zone, and 7 percent of the humid zone are set on fire annually, impeding the much-needed cycling of nutrients back into the soil.

Africa therefore challenges the intuitive belief that environmental destruction from farming must be the result of producing too much food or using modern high-yield technologies. Africa has not yet embraced such technologies and is producing too little food rather than too much, yet it is still degrading its rural environment. The environmental resources that will be needed badly by tomorrow's farmers are, in too many cases and for too little advantage, being destroyed by the farmers of today. Africa's problems are of the first-generation variety. Most are the result either of extending traditional low-yield, low-input farming into forests and onto new lands or of cropping the same land more often without replacing nutrients (nutrient mining). Rather than suffering from excessive use of farm inputs, Africa's rural environment is suffering from too little use of inputs. Fertilizer use in Africa, at 12 kilograms per hectare, is only about one-quarter the level of India and just one-thirty-sixth the level of Japan. Irrigation covers only 4 percent of cultivated area in Sub-Saharan Africa, compared with 26 percent in India and 44 percent in China.

If Africa does not soon boost yields on its existing farmland, the rest of its rural environment will come under intensifying threat. Rates of increase in average yields will have to more than double from current trends if a 4 percent growth in production is to be achieved without speeding up the current rate at which nutrients are mined or area is expanded.[35] If agriculture is not intensified, many actions that would otherwise be desirable—such as giving higher price incentives to farmers, building more rural roads, reducing the risks to human and animal health, or relieving seasonal labor shortages—could actually become environmentally undesirable. As more roads are built into remote areas (for example, logging roads into forests) and as medical advances eventually reduce the vulnerability of humans and animals to site-specific diseases (for example, trypanosomiasis, or sleeping sickness), farmers who have not increased production on the lands they now crop or graze will move into still more fragile areas and ruin them in the process.

Some environmentalists fearful of the second-generation problems that can accompany green revolution farming have been slow both to accept this modernization and intensification imperative and to endorse the large investments in nontraditional practices, such as irrigation, or in nontraditional purchased inputs, such as fertilizer or hybrid seeds, required to make Africa's system of food production environmentally sustainable. Given the reality of rapid population growth, however, Africa has no practical alternatives. Sub-Saharan Africa's population is now doubling every two decades. Even if fertility rates could be reduced 50 percent between now and 2030, Africa's population would still grow from 494 million persons in 1990 to 1.2 billion by 2020 (by 2075 the region could be more heavily populated than Asia). Meanwhile, if Africa's current 2 percent annual growth of food production cannot be accelerated, the deficit between production and consumption of food could increase more than sevenfold between 1990 and 2020, from 10 million to 74 million tons of food grain equivalent.[36]

Significant gains in medicine and public health, especially inoculations against disease, have cut infant mortality rates and helped bring on Africa's exceptionally high rate of population growth. Birth control alone cannot halt Africa's population growth in less than a generation's time, since the continent's next generation of parents has already been born. Africa's AIDS epidemic will slow population growth, tragically, but mostly by increasing the mortality of adults who have already had children. The disease will contribute much more to human misery and to burdensome public health expenditures than to population stability.[37]

Could Africa perhaps avoid intensifying its farming sector by increasing its commercial food imports (following the path proposed above for Asia) or by importing still more food aid? Commercial food imports are not affordable for Africa, and donors might not be willing to supply aid on the scale needed. Also, food imports or aid might benefit the one-third of Africa's people who live in urban areas and who receive and consume the food in areas adjacent to the pier. However, they might harm the two-thirds who live in rural areas and who may not be targeted to receive concessional food imports. These persons may even suffer a decline in the price of the agricultural goods they produce and sell.

Clearly, Africa cannot escape the need to invest more in farming itself. In contrast to the rapidly industrializing parts of Asia, Africa's overall economic growth remains heavily dependent on agricultural prosperity. An average of 32 percent of gross domestic product (GDP) is still produced by agriculture (compared with just 11 percent in Korea). Also, agriculture is the dominant contributor to growth in manufacturing and services; it is the largest source of raw materials for industry and the main purchaser of simple tools and services; and farmers are the major purchasers of consumption goods. Boosting the productivity and prosperity of African agriculture through green revolution investments in water, seeds, and fertilizers is essential to achieving the larger goal of economic development beyond the farm sector.

In taking these necessary steps to make farming more productive, Africans face unique difficulties. Particularly compared with East Asia, Africa is endowed with more difficult climate and poorer natural resources for intensified farming; its demographic transition is more sudden and more problematic; and its political institutions are less stable, more centralized, and less accountable.

NATURAL DISADVANTAGES

Africa's natural resource base for agriculture is on the whole less productive, less durable, and more hostile to farming than that of either Asia or Latin America. Only 6 percent of the continent is used for any kind of arable crop.

- Soils tend to be poor even by the standards of a tropical country: They are highly weathered, acidic, and low in fertility.
- Rainfall tends to be scarce, excessive, excessively violent, or unreliable; an estimated two-thirds of the continent is subject to high risk of drought, and some 46 percent has less than 75 days of rain a year, too little even to raise millet. Tree planting, normally an option for conserving the soil, is not viable in large parts of Africa, which receive less than 1,000 millimeters of rainfall a year. Because water is uncertain, farmers spend much of their time looking for ways to reduce risks; water and weather risks reduce the attraction of buying inputs to increase yield.
- Temperatures tend to be too high, causing excessive evaporation in semi-arid zones and depleting soil organic matter in deforested humid zones. Africa's farming regions are seldom variable enough in altitude (except in parts of East Africa) to provide relief from the sea-level heat or to sustain a wide mix of crops.

- Local topography can be highly irregular, complicating irrigation engineering (and the construction of rural roads). Irrigation costs in Africa are roughly double those of other continents. Most of the potential for added irrigation is distributed unevenly in just four large countries (Angola, Mozambique, Zaire, and Zambia), none of which suffers yet from a shortage of rainfed land. Meanwhile, the countries of the Sahel, which desperately need more farming land, have scant additional potential.
- A wide variety of pests and diseases routinely attack crops, livestock, and people. As much as one-third of tropical Africa remains unexploited because of the presence of trypanosomiasis, which affects both people and livestock.[38]

There are limits to what green revolution plant breeders can provide under such circumstances. The most important achievements in plant breeding have traditionally come from increasing the ratio of grain to straw, rather than from increasing biomass itself. Options that are available in the wet regions of Asia—such as the engineering of plant varieties that yield more grain by taking more water and nutrients out of the environment—are simply less available in Africa, because the local supply of water and nutrients is less abundant. Efforts to introduce highly responsive seed varieties into Africa, even for crops that are already being grown, such as sorghum, millet, and rice, have met repeated failure.[39] Laboratory science can be applied profitably to African farming, as in the development of crop and animal varieties that tolerate droughts or weeds, but the resulting benefit usually consists of minimizing downsize losses to farmers rather than of offering immediate, significant upside gains.

DIFFICULT DEMOGRAPHICS

Africa's more problematic demographic transition also inhibits farm intensification. Ordinarily, population growth should stimulate intensification, because it reduces the cost of labor relative to land, enhancing the substitution of labor for land (as through mulching, the application of nutrients, stall feeding of livestock, or the construction and maintenance of terraces and systems for harvesting water). Rural population density also makes affordable the key infrastructural investments, such as farm-to-market roads and irrigation systems, typically needed to support intensive styles of farming requiring access to purchased inputs and commercial markets. Gradual increases in population density also tend to induce a long-term evolution in social and legal resource control institutions away from open-access regimes, toward more secure common property systems, and eventually toward individual private property systems. As control over land or trees becomes more secure in this fashion, farmers have more incentives to invest in the long-term productivity of land, with nutrient replacement, irrigation, or agroforestry.

These population-linked inducements to intensification are widely visible in Asia, but they have yet to emerge in most of Africa, for at least three reasons. First, despite unprecedented annual rates of increase, rural populations in most of Africa are still too sparse to justify large substitutions of labor for land or to make public investments in roads or irrigation seem affordable. Compared with Asia, Africa still has 4.5 times as much total land in arable and permanent crops and pasture for each economically active person engaged in agriculture. Partly because so many men have left villages to seek work in towns and cities, labor *shortages* are actually a problem in many rural areas.[40] In some parts of central Africa, the AIDS epidemic could soon worsen these rural labor shortages.

Second, the historical evolution of land control institutions has been badly disrupted by extreme swings in the location and density of population. Unlike Asia, Africa suffered a significant *depopulation* up through the eighteenth century under the impact of the slave trade. In the trans-Atlantic slave trade alone, roughly 8 to 12 million Africans were sold into slavery before 1820, and many more died or were killed by slave traders before they could make the journey.[41] Then, quite abruptly in the mid-twentieth century, Africa swung to a new pattern of extremely rapid increase in population, as modern public health, anti-malarial, and vaccination procedures suddenly became available in rural areas. Whereas Asia had centuries in which appropriate common property and private property institutions evolved slowly in pace with population, in rural Africa many traditional institutions of land control have reverted back to environmentally damaging open-access regimes because population is increasing at such a high rate.

The evolution of appropriate local institutions is also blocked by pervasive public expropriation and ownership. Local farming and foraging communities originally lost control over their lands and forests to foreign colonizers. Decades have passed since most of Africa gained its formal independence from European rule, but lands that were lost in many cases remain alienated from local control, often under state ownership. In Angola, Kenya, Mozambique, Swaziland, Zambia, Zimbabwe, and elsewhere, many Africans were displaced from good farmlands to make way for white settlers, and to this day many remain crowded onto small areas of only marginal agricultural potential (ironically called "homelands" in South Africa). Government land titling programs are often operated as patronage schemes, designed to grant urban elites access to exploitable rural resources. Such schemes too often threaten the access rights of holders who do not have title and increase insecurity and landlessness.[42]

POOR GOVERNANCE

Overly centralized, urban-biased, and nonaccountable governments are a third barrier to the timely intensification of farming in Africa. These severe problems of governance date from colonial times, when political power was conferred on an alien class of coastal or city-dwelling European administrators after coastal ports of embarkation replaced interior trade routes as the dominant paths of commerce. When the colonizers departed, they typically handed over highly centralized instruments of taxation and control of agricultural markets to a new class of urban-based African political leaders, who were often hostile to agriculture because of its colonial associations and who were intent on rapid development of urban industry. The result was an early and persistent underinvestment in farming, an overtaxation of rural communities to generate resources for urban investment and consumption, plus an overly centralized (mis)management of rural resources.

Since independence, most governments have continued to take from the farm sector and give back as little as possible in return. Agriculture accounts directly or indirectly for 80 percent of employment in Africa and for 50 to 90 percent of export earnings, but national governments typically devote less than 10 percent of their public spending to the farm sector. In the 1980s, national expenditures on agriculture actually fell in many countries in Africa, and a larger share of what was spent came to be financed through foreign aid.[43] In 1985, governments within the Organization of African Unity pledged to increase their share of agricultural spending to 20–25 percent, but the pledge was not fulfilled. In Cameroon, the government recently spent more on the construction of a single airport in the capital city than on its entire nationwide agricultural development effort.

Since independence, macroeconomic policies have also damaged African agriculture. Governments use overvalued currency exchange rates to cheapen the purchase of imported consumer goods, including food. This, when combined with inadequate public investment in all-season farm-to-market roads, reduces commercial opportunities for Africa's own producers. In many large coastal cities, it is cheaper to purchase food from abroad than from the nation's own disconnected farmers living nearby.

African governments also hold back the performance of agriculture by maintaining costly and inefficient public ownership of the supply of inputs, marketing, irrigation, and production. These government parastatal organizations frequently own the largest tree crop plantations, cattle ranches, agricultural marketing enterprises, irrigation facilities, and crop processing factories. In the vast majority of cases, these publicly owned or managed operations perform less well than competitors in the private sector.

For more than a decade, international financial institutions such as the World Bank and the International Monetary Fund (IMF) have extended structural adjustment loans to African governments in an effort to reform such macroeconomic and sectoral policies. Some countries have embraced reforms (Ghana is now the leading example), but many have not. A 1994 World Bank review of 29 Sub-Saharan African countries found that during the 1980s none had achieved what the World Bank considers to be a sound macroeconomic stance, and macroeconomic policies in one-third of them actually deteriorated. In the agricultural sector specifically, 17 countries managed to reduce the overall tax burden on farming, but some—because of persistently overvalued exchange rates—actually increased it. Only four of the 29 countries eliminated parastatal marketing boards for major export crops. This study drew the conclusion that "no country has good macroeconomic policies *and* good agricultural policies."[44]

Direct government environmental protection policies in rural areas are also weakened by overcentralization and lack of participation. Soil and water projects, developed and implemented by central government ministries and foreign technical advisors without adequate commitment and participation at the level of the farm or village (much as in colonial times), are seldom successful. The village-level resource-protection strategies that once existed in Africa, including traditional common forest or rangeland management systems, are often broken apart and replaced by centrally directed government expropriation.

Africa's governments seek to maintain their centralization and urban bias by intentionally blocking the emergence of powerful local institutions in rural areas. Africa's popular grass-roots political organizations are weaker than those of either Asia or Latin America. There are numerous social clubs, secret societies, and religious sects at the local level, but the state tends to monopolize local politics as well as the operation of the local economy. The most important instruments of local agricultural control—genuinely independent farm cooperatives—are typically missing in Africa. Local control is also missing in the public sector; local government employees constitute just 10 percent of all government employees, only half the ratio for most countries in Asia or Latin America.[45]

To date, governments have done too little to help Africa's women farmers. Even Kenya, one of the region's most progressive states in rural social and agricultural policy, has not yet found adequate means to reach women farmers with technical and credit assistance. Kenya's agriculture is increasingly in the hands of women (roughly two-fifths of Kenya's rural households are headed by women), yet most of the government's 3,500 frontline extension agents are men, and very few women smallholders enjoy access to Kenya's formal rural credit institutions. Women in rural areas represent no more than 10 percent of all formal loan recipients. Kenya's government-funded Agricultural Finance Corporation lends to about 20,000 farmers, of which hardly any are women. Of 2,549 registered farm cooperatives, only 28 were registered as women's cooperatives.[46]

At the national level, Africa's political institutions are seldom democratically accountable. A recent inventory of 46 governments found that only five of the smallest could be classified as genuine multi-party systems.[47] In more than half, opposition parties were not even legal. A number of elections did occur in Africa early in the 1990s, but most were controlled by former one-party rulers, who in many cases managed to arrange their own victory. Of more than 150 national leaders in post-independence Africa up through 1990, only six left office voluntarily.

One final characteristic of African politics that inhibits the intensification of farming is the frequency of armed conflict. Because African governments are so reluctant to share power peacefully, they often invite armed challenge. The difficult period of decolonization in Africa was marked by at least 10 major post-colonial wars, and many of those conflicts are still active today. At one point in 1993, a 17-year-old civil war was still under way in Angola; a three-sided military conflict continued in Liberia; a tenuous peace (following a disastrous 16-year civil war) was only beginning to take hold in Mozambique; a long-standing civil war between the Hutus and the Tutsi was again raging in Rwanda; more than 230,000 refugees had fled fighting in Togo; rebels in the mostly Christian and animist south of Sudan were still fighting the Muslim-dominated government (and each other as well); and U.S. military forces had just arrived in Somalia to protect food relief shipments from attacks by the armed clans that had two years earlier overthrown the government of President Siad Barre.

Sub-Saharan Africa recently had a per capita war fatality rate three times as high as the Middle East. Over the last 30 years, 7 million Africans have died fighting, and millions more have died due to accompanying disruptions of food markets and basic social services. In mid-1994, even before the tragic Rwandan exodus into Zaire, more than 6 million people were refugees, and 16 million more were "internally displaced persons" (about three-fifths of the world total).[48] Economic growth has not fared well amid this violence and dislocation, and neither have resource protection or farm intensification.

EXAMPLES OF INTENSIFICATION

■ IN THE FACE OF SUCH DAUNTING CONDITIONS, in some cases, African agriculture has been successfully intensified, and the environmental gains are clear. In Kenya, during the 1970s when governmental policies were less predatory toward the farming sector than they are today, farming was substantially intensified on high-potential lands, especially in the central highlands. Between 1972 and 1988, Kenya's total area under certified maize seeds increased threefold, and total fertilizer consumption increased at annual rates above 10 percent. By 1977, high-yield varieties of hybrid maize, which gave yields as much as 40 percent higher than traditional varieties, had been adopted by half of all Kenya's smallholders. Terracing, designed both to harvest water and to protect soil, boosted crop yields an estimated 30–65 percent. This successful process of intensification, as expected, reduced environmental stress. In the Machakos District, according to a 1991 study conducted in part by Great Britain's Overseas Development Institute, "a favorable environment for attaining sustainable agriculture" had finally been created, and land degradation, which had been widespread, was "reversed."[49]

In Kenya today, some of this farm intensification is faltering, in part because governmental policies are less supportive of it. Small coffee growers have lost the right to form a union; government corruption has hampered delivery of fertilizers and other inputs; and crop marketing boards have been changed from instruments of farm support into instruments of central control and taxation. Annual growth of the farm sector fell from an average of 5 percent before 1980 to 3.3 percent in the 1980s; production of coffee, maize, sugar, and wheat stopped growing and began to *decline* in the early 1990s. Fertilizer use peaked in 1988–89 and subsequently declined. High-value coffee trees are being neglected, and investments in soil protection are being forgone, even though Kenya's population is just now entering its period of most dramatic projected increase.[50]

In Nigeria, the largest country in Africa, intensification has also taken place, albeit with fewer environmental gains. During the 1970s, due to high oil prices, the government in Nigeria was relieved of any need to continue extracting revenues from its faltering farm sector. It halted the taxation of farming, and even provided subsidies. During the oil boom, the government spent billions of dollars on the construction of rural roads, generous subsidies for farm inputs (including an 85 percent subsidy on fertilizer), and a variety of costly agricultural development projects. Between 1970 and 1982, total farm production grew 2.7 percent annually, and food production grew even faster, actually keeping pace with population growth.[51]

Unfortunately, much of this intensification was of the wrong kind and in the wrong place. Instead of intensifying farming in the more densely populated southern parts of the country, where environmentally sustainable farming of cocoa and oil palm tree crops was an attractive option, the Nigerian government sought mostly to introduce high-input farming of cereals into some of the less densely populated regions of the north as a means to replace food imports for the urban population. Nigeria's "maize boom," induced in the northern savanna area with excessive fertilizer subsidies, has proved to be a mixed environmental blessing. Maize production has reduced both organic matter and micronutrients in the soil, increased the incidence of weed pressure, and aggravated soil acidity.[52] Nigeria's larger irrigation schemes and its river basin development projects, designed for mechanized wheat and rice cultivation, have also generated severe ecological problems.

SUMMARY PRESCRIPTIONS

■ NEITHER AFRICANS NOR OUTSIDERS have yet taken a responsible posture toward satisfying the region's looming food production needs or addressing its serious demographic and rural environmental crisis.[53] To avoid a slow-motion environmental and Malthusian nightmare, larger investments must be made in rural roads, fertilizer, irrigation, extension, and research. Policies and political institutions must be reformed to give farmers and local farm communities greater control, improved participation, and stronger incentives to make resource-conserving investments on their farms. All this requires revolutionary changes in the attitudes of urban politicians toward rural communities and even in the attitudes of men toward women. Education and employment for women, plus family planning and maternal health care, are essential elements in the effort to slow the damaging momentum of Africa's population growth.

In Africa, partly because population growth rates are so high, rural environmental damage often takes the form of expanding traditional farming practices onto fragile lands. To protect these lands while increasing the supply of food, it is necessary to increase yields on farmland now in use by applying more water and nutrients, including inorganic fertilizers. The dry parts of Africa can never intensify as much as the wet parts of Asia, but a faster step-by-step movement in the direction of higher yields is now a matter of life or death, both to protect human nutrition and to save the rural environment. Africa therefore needs many of the same investments in fertilizer, irrigation, seeds, roads, research, extension, and family planning that are needed in the less-favored areas of Asia.

In addition, because so many African farmers are women, the continent needs far greater gender equity in rural social relations. Because African governments too often smother the private sector, progress toward liberalization, privatization, decentralization, and macroeconomic adjustment must continue. And the frequent military struggles in rural areas that make profitable and environmentally sustainable farming impossible must also, in the interests of all, be brought under control.

Solutions to Africa's problems can be divided into short-, medium-, and long-term approaches. In an excellent review of Africa's agricultural development and environmental protection challenge, Kevin M. Cleaver lists four actions that could have an immediate impact within one year: change in agricultural policy, agricultural extension, food security and nutrition initiatives, and farm input supply initiatives. At least a dozen actions could have a constructive medium-term impact within one to five years: road building, soil conservation, development of farmers' groups and cooperatives, proactive development of the private sector, irrigation development, credit, privatization, regional integration, improved governmental management, reform of land tenure and forest policy, and agricultural research. Finally, eight actions could be undertaken with long-term advantages: education, population policy, health services, biodiversity conservation and national environmental actions plans, water resource and land management, drought contingency planning, geographic information systems, and sound urban policies.[54] Most African governments have been slow to respond to these massive challenges, yet outsiders have no grounds to complain so long as their own efforts remain so tentative and so small. The foreign assistance community should make larger investments in food production and environmental research, should support the more expensive means of data collection (such as aerial photography and remote sensing), and most of all should make long-term investments in education, family planning, training, rural health, and institutional capacity-building. Nobody (Africans or outsiders, governments, international organizations, or NGOs) has an excuse for inaction.

Part IV
Latin America

DUALISTIC FARMING, DUALISTIC RESOURCE ABUSE

LATIN AMERICA IS WELL ENDOWED BY NATURE for productive and environmentally sustainable agriculture. From the sunny and irrigated lands of northern Mexico to the humid pampa of Argentina, high-quality natural resources for farming are relatively abundant. Rural population densities tend to be low, outside of Central America, and the demographic transition tends to be more advanced than in Africa or Asia. Moreover, Latin America is relatively well off: In 1989, gross national product (GNP) per capita stood at nearly $2,000, almost two-thirds higher than the average for East Asia, five times higher than South Asia, and six times higher than Sub-Saharan Africa. Despite these advantages, Latin American farming remains a source of serious rural environmental destruction.

In Latin America's agricultural sector, there is a curious mix of both first-generation and second-generation environmental problems. Large commercial farms use or mismanage too many water, chemical, and machinery inputs, often side-by-side with small peasant farms that damage the land and forests by failing to modernize. This mix cannot easily be attributed to nature, to demography, or to poverty. Nor can it easily be addressed with macroeconomic reforms, technical breakthroughs, or new investments. Too many of the environmental problems facing farming in the region reflect deeper problems of social inequity and marginalization.

HISTORICAL FOUNDATIONS OF DUALISM

Latin America's social history is less fortuitous than its natural resource endowment. The arrival of conquering Spanish armies early in the sixteenth century placed a permanently damaging mark on the society and agriculture of the region. The original European conquerors mostly came in search of gold and silver, yet control over the land was important to their Iberian culture, if only for reasons of social status, and the tragic depopulation of tropical America facilitated their acquisition of land, including good farming land. An estimated three-quarters of the indigenous peoples of the Americas south of the Río Grande were eliminated between 1492 and 1650, due to a combination of brutal treatment and deadly exposure to unfamiliar European diseases such as smallpox, measles, and typhus. The conquerors established vast *hacienda* estates, which grew crops and raised cattle for the still-small urban areas where most Europeans (including estate owners) preferred to settle. Peasants were permitted to farm small plots of land for subsistence purposes, in return for offering their labor to the landlord. They were tied to the estate in a semi-feudal fashion, partly through their debts, which accumulated across generations, and partly through their vulnerability to local legal institutions and police powers, which were dominated by landlords. Indigenous peoples who were not incorporated into the *hacienda* system were either resettled into crowded Indian towns on inferior land or were pushed into remote forest and upland areas.[55]

This system of highly inequitable control over both land and labor still drives much of the region's abuse of rural farm resources, despite significant changes in land use over the years. Beginning in the late nineteenth century, modern international transport technologies plus industrial development in Europe and North America created a lucrative international market for the agricultural commodities of the region, including bananas, coffee, cotton, rubber, sugar, and tobacco. Latin American political elites (now independent from Spain or Portugal) saw an advantage in transforming more of their farming systems into export crop enterprises, often with financing from foreign banks and sometimes even with foreign ownership. Good farming land gained value during this export boom, so landowners demanded higher payments and rents from sharecropping peasants and tenants, squeezing many of them off the *haciendas*.

Some governments also embraced supposedly "liberal" land market reforms during this period, allowing traditional peasant and Indian lands to be freely bought and sold, but wealthy ranchers and land speculators could always outbid the poor. By the early twentieth century, as a consequence, concentrations of landownership reached new extremes in most of Latin America. In Brazil, by 1920, just 4 percent of all property holders controlled 63 percent of total landholdings. In Chile, by 1924, 3 percent of all property holders controlled 79 percent of total landholdings. The landowning elite that profited from the export boom spent much of its time and money pursuing an affluent urban lifestyle, further widening the economic and cultural gap between city and countryside.[56]

The elites reconsidered their heavy dependence on farm exports after the Great Depression of the 1930s led to a catastrophic collapse in commodity prices, and as their own evolving urban tastes made them impatient for industrial development. Especially after 1949, a new import-substitution strategy for industrial development was widely promoted by Raúl Prebisch and other economists at the newly created United Nations Economic Commission for Latin America. This strategy implied that wealth would have to be squeezed out of the agricultural sector, but it also meant that farm production would have to be improved and modernized enough to be capable of generating more taxable wealth. The squeeze was applied by creating and manipulating monopoly government marketing boards, by embracing overvalued exchange rates and high border protection for industry, and by directing only a small share of government spending (typically less than 5 percent) back into the farm sector. To preserve adequate wealth in farming (so as not to kill the goose that was laying the golden eggs), the minority of potentially productive commercial farms received offsetting benefits. A narrowly based commercial farming elite was given subsidized credits, research and extension assistance, improved storage facilities and transportation networks, new irrigation canals, exemptions from import duties on farm tractors, and subsidies on fertilizer, pesticides, and new seeds.

This strategy increased levels of production, but without improving the fortunes of ordinary farmers and without protecting the rural environment. Between 1955 and 1974, agricultural production grew at an average annual rate of 2.9 percent, higher than either the U.S. or the Canadian growth rate for that same period. Between 1960 and 1974, the dollar value of agricultural exports from the region actually tripled.[57] This spurt of production was strong, but hardly an Asian-style green revolution. Too many of the gains in production were based on area expansion (often encouraged by mechanization) rather than on higher yields. The environmental costs included the cutting of more trees and the plowing of more marginal lands. Between 1961 and 1991, 30 percent of the total increase in cereals production was a result of expanding the area under cultivation; in East Asia, by contrast, only 6 percent of the (much larger) total increase in cereals production was attributable to area expansion.

Also, when intensification strategies were pursued in Latin America, they were too often based on a narrow use of subsidized inputs, such as irrigation water and chemical pesticides, by only a small privileged minority of commercial farmers. This left the rural majority, which received no intensification support, unable to purchase fertilizer, to incorporate livestock into their operations, or to defer income long enough to farm in rotations. As a consequence, they continued to exhaust soils with low-input farming techniques, then move onto even more fragile land or into the forest margin. In contrast to Asia, where the switch to science-based farming usually helped ordinary peasant farmers, the modernization of farming in Latin America frequently hurt the poor, who then hurt the rural environment.

These links between agricultural dualism and environmental destruction are so strong in Latin America that remedies based on superficial policy or technical changes alone produce only limited gains. Consider two case studies: the limited gains that have come from policy reforms in Mexico and the limited gains that have come from technical change (the introduction of high-value crop farming) in Central America.

MEXICO: LIMITED ENVIRONMENTAL GAINS FROM STRUCTURAL ADJUSTMENT

Despite a history and ideology of social revolution, Mexico continues to exhibit rural social inequities almost as extreme as those found in other Latin American nations. Farmers tend to be either relatively affluent or very poor. A small elite owns the best irrigated lands and has profited over the years from generous government subsidies, while the vast majority of farmers are confined to lands with little water, poor soils, and disadvantageous terrain. Roughly half of Mexico's farmers are trying to subsist on only 10 percent of the nation's farmland, and an additional number of rural laborers—perhaps 600,000–700,000 in all—are entirely landless.[58]

Both rich and poor farmers are engaged in damaging forms of environmental abuse. The poor tend to abuse the environment by invading forests and fragile lands, by failing to add fertilizer to replace nutrients, and by making few investments in soil protection (such as contouring, tree planting, or fence building). They often lack both the means and the incentive to protect their farms, because their land is so poor to begin with (they spend most of their time looking for income opportunities off the land) and because they often lack secure title (they are legally marginalized or work communally owned *ejido* lands). Fearing displacement, they have little incentive to protect the land. The result is first-generation environmental damage on a large scale. Since 1961, Mexico has lost about 36 percent of its total forested area, and more than half of the soil area is either totally eroded or undergoing accelerated erosion. Large parts of rural Mexico resemble a lifeless moonscape. In the Mixteca region, 70 percent of the potentially arable land is no longer able to grow crops due to soil erosion.[59]

Meanwhile, on Mexico's high-potential irrigated land, the elite of commercial farmers engages in an entirely different kind of environmental abuse. As a result of state subsidies for fertilizer and pesticides (which are at last being cut), large growers engage in excessive use of chemicals. Use of nitrogen fertilizer per hectare increased 800 percent (from a low level) between 1961 and 1989, and during the decade of the 1970s, use of pesticides increased at an average annual rate of 8.4 percent. In the Culiacán Valley, tomato growers spray pesticides on their crops as often as 25 to 50 times per growing season. Such practices endanger the downstream environment (the Culiacán River leads to a series of lagoons where migrating wildfowl spend the winter and where local shrimp fishermen complain of a falling catch) and harm the farmworkers who transport, mix, and apply the chemicals. Farmworkers in Mexico are willing to undertake this dangerous work because they have no other way to earn the seasonal income they desperately need. They still have few legal rights to organize, and many of the workers are children.

Policies in Mexico are changing rapidly, but so far with limited effect on the deep structure of social dualism in the countryside. Under the government of President Carlos Salinas de Gortari (1988–1994), Mexico undertook significant improvements in macroeconomic discipline, widespread privatization (including privatization of communal farmlands), market deregulation and free trade under the North American Free Trade

Agreement (NAFTA), and more ambitious efforts to target social spending toward poor people, under the National Solidarity Program. Unfortunately, these reforms go only part way toward slowing environmental abuse in agriculture.

Mexico's commitment to greater budget discipline—a necessary step to securing its 1989 debt agreement with the IMF—has cut subsidies for inputs to commercial farmers, which reduces the incentives to irrigate, mechanize, fertilize, and spray excessively. The state budget has been cut overall, and the share of spending going to support expensive, large-scale, capital-intensive irrigation projects—mostly in the rich north—has also been reduced.[60] However, Mexico's adoption of agricultural free trade under NAFTA could have negative as well as positive environmental consequences. As the sale of winter vegetables from Mexico to the United States expands under NAFTA (an anticipated 40–50 percent increase between 1990 and 1998), the amount of acreage devoted to pesticide-dependent tomatoes could increase in irrigated districts such as Sonora.[61]

Probably the most important of the Salinas reforms, for both social and environmental purposes, is the 1991 initiative (approved early in 1992) to amend article 27 of the constitution to permit the private sale or rental of communal *ejido* lands. In theory, this change could have positive effects on the environment. If small peasant farmers are allowed to own their lands individually, they may finally gain both the incentive and the collateral needed to finance resource-protecting investments. Some of these lands might also be better protected from misuse if they were sold or rented to full-time commercial farmers. Nevertheless, small peasant farmers have seldom been able to use free markets to their advantage in rural Mexico. In the spring of 1994, the U.S. Embassy in Mexico City described as "socially catastrophic" the foreclosures on rural property that were already taking place among ordinary peasants due in part to the Salinas reforms, especially the privatization of rural credit. Inequitable privatizations could trigger a large exodus of labor from rural areas at a time when the rest of the Mexican economy is not producing enough jobs. Some new farming jobs will be created in the expanding fruit and vegetable sector, thanks in part to NAFTA, but even the most optimistic observers do not see NAFTA adding more than about 100,000 new jobs annually in Mexico's entire economy.[62]

It has not helped that Mexico has made these sweeping changes in policy in a highly centralized, top-down fashion, with poor peasant farmers feeling little sense of control over the outcome. Participation was to be a part of President Salinas's National Solidarity Program (PRONASOL), which was built around roughly 150,000 local leadership councils designed to deliver government services and support directly to the poor, outside of regular government bureaucratic structures. This program did little to empower poor rural farming communities. Especially in southern Mexico, where many poor farmers are indigenous peoples facing racial discrimination as well as social marginalization, the Solidarity approach too often was blocked by local political bosses or taken over by top-down government authorities.[63]

The continuing inadequacy of these reforms, especially from the vantage point of poor indigenous farmers in southern Mexico, was expressed dramatically in the January 1994 uprising of the Zapatista National Liberation Army in Chiapas, an area of extreme social injustice and severe environmental degradation. Government authorities, not wishing to undertake serious land reform in Chiapas, had attempted to move landless families into the jungle, where they cut the forest to raise coffee, cattle, and subsistence crops. The cleared land quickly lost its fertility, and often these farmers were pushed still deeper into the forest by cattle ranchers with political ties to local elites. The local Chiapas elite, openly racist in its views toward the indigenous Mayan farmers, systematically denied them the secure control over resources that is essential to promoting environmental stewardship.[64]

Although the government's initial public response to the uprising was hard-line, a combination of international and domestic pressures for a peaceful solution quickly moved Salinas to declare a unilateral ceasefire, open constructive talks with the insurgents, and agree to independent oversight of election commissions, independent auditing of voter rolls, fair access to the media, and lower ceilings on campaign spending—measures that could free some of Mexico's currently "controlled" rural voting districts, such as Chiapas, to elect their own authentic representatives and break the power of the traditional, conservative elite. Nonetheless, when Mexico's 1994 elections were finally held, the governorship of Chiapas state was won by the Institutional Revolutionary Party, or PRI (the governing party), amid opposition allegations of widespread fraud. Too often in Latin America, democratic gains are strictly an urban middle-class phenomenon, with rural areas remaining under the control of traditional elites.[65]

CENTRAL AMERICA: LIMITED GAINS FROM TECHNICAL CHANGE

Whereas Mexico illustrates the limited impact of privatization, trade liberalization, and conventional macroeconomic reform in the absence of political change, Central America illustrates the limits of technical change. Central America is a relatively well-endowed agricultural region that has been able, over the years, to adopt important technical innovations in the production and marketing of commercial farm products, including high-value products for export to nearby markets in North America. After World War II, export booms in cotton and cattle brought considerable wealth to the region; recently, Central America has been experiencing an export boom in nontraditional fruits and vegetables (such as off-season blackberries, broccoli, raspberries, snow peas, squash, and strawberries). Movement toward higher value farm production can be good for the rural environment, as we saw in parts of Asia and Africa, because it allows farmers to substitute the use of labor and capital inputs for the use of forest lands, erodible lands, and scarce water. Because of inequitable rural social and political institutions, however, this sort of environmental advantage is much less likely to be gained in Central America.

Farm modernization in Central America has frequently brought the unhappy combination of commercial success alongside social and environmental failure. Modern farming entered the region after World War II, primarily as a result of the construction of roads (the Inter-American Highway system, begun during the war, and the Pacific Coastal Highway, promoted beginning in the 1950s by the World Bank) plus the development of chemical insecticides, such as DDT, which made the low-cost production of crops such as cotton a technical possibility.[66] This sudden introduction of profitable commercial farming into what had been static, semi-feudal rural societies rapidly created wealth, but only for a few.

Most of the rural poor failed to participate in the new commercial farming operations, and they and their lands increasingly fell victim to those who did. As a consequence, the mismanagement of farm inputs increased among the newly rich, while land insecurity, dispossession, and resource abuse increased among the rural poor.

The biggest commercial opportunities were seized by urban-based agricultural entrepreneurs who had both the capital resources and the technical skills needed to rent land and shift it into the production of higher value crops such as cotton and cattle. Since these new operators often used land on a one-year cash rental basis, they had little incentive to maintain soil nutrients or prevent soil erosion over the long term. They also pursued short-sighted strategies that sought the "total eradication" of pests, making Central America, on a per capita basis, the world's heaviest user of pesticides.

As more traditional lands were shifted into commercial cultivation of cotton, or into the grazing of beef cattle, ordinary peasant farmers lost their traditional access to land for producing food crops. Access rights on private *haciendas* were switched from traditional forms (sharecropping or in-kind rent) to money rents paid in advance, which peasants could not afford. The evicted peasants gained some compensation in the form of seasonal employment two or three months a year as hired cotton pickers but otherwise either became slum dwellers on the fringes of the urban economy or tried to grow subsistence crops on lands that were vacant because they were not suited to mechanized commercial agriculture. In practice, this often meant drier lands, forest lands, sloping lands, or lands with shallow and less fertile soils, all of which were subject to rapid degradation.

As many as 3 million people in Central America—roughly one-fifth of all rural inhabitants—have recently been displaced by commercial eviction, political intimidation, or war.[67] Meanwhile, regions that could at one time absorb new settlers have been deforested, depleted, and environmentally destroyed. More than 50 million metric tons of soil are lost a year. The cost of replacing the nutrients contained in this quantity of lost soil, at current market prices, exceeds the annual budget of all five countries in the region. An estimated 250,000 hectares of forest are lost annually, largely the result of slash-and-burn farming practices. By one calculation, Costa Rica depreciated its soils, forests, and fisheries by more than $4.1 billion between 1970 and 1989, a sum exceeding the average value of one year's GDP at the time.[68]

A particularly difficult case is Honduras, where the post-World War II cotton boom led to even higher concentrations of landownership and control, followed by widespread peasant evictions and large-scale rural environmental damage. As a result of the cotton boom, the largest 4 percent of all farms eventually controlled roughly 55 percent of the nation's farming area, including most of the high-quality lands. Roughly two-thirds of all the remaining farms and farmers were left sharing less than 10 percent of the total farming area, including the most fragile and least productive lands. The burden on these fragile lands was all the while being magnified by rapid population growth, which resulted from the introduction of rudimentary medical and public health services into rural areas; the total population of Honduras roughly doubled between 1970 and 1989, forcing ever larger numbers of destitute subsistence farmers into forest margins or onto steep hillsides. As a consequence, fallow land has declined precipitously, soils are being lost at a rate of 10,000 hectares a year, and forest resources are at risk of being totally exhausted, perhaps within a generation.[69] Nevertheless, many of the best and most durable lands in Honduras remain underused by absentee owners who often keep these lands in pasture.

Without deep social reforms, these problems will persist. Landownership is badly in need of reform in Central America and throughout the rest of Latin America; even some of the land reforms implemented under "revolutionary" circumstances in both El Salvador and Nicaragua in the 1980s fell short of reaching a majority of the rural poor. Barring a return of revolutionary circumstances, probably the best hope is a modest program of tax or subsidy reform to reduce some of the benefits (low land taxes, input subsidies, and subsidized credits) that have tended to inflate property values for large landowners. Lowering property values could eventually make the purchase of good farming land more affordable, either by small farmers directly or by reformist governments on behalf of small farmers.[70] Raising property taxes, especially if the revenues were controlled locally, could deflate land values and raise revenues for investments to protect local resources at the same time. Yet even such modest steps will require more political courage than has recently been evident in Central America.

Short of such reforms, can technical change alone provide an answer? Consider Central America's latest commercial boom in nontraditional fruits and vegetables. Between 1983 and 1990, Costa Rica's exports of fruits and vegetables to the United States (*excluding* bananas) increased at an astonishing rate of 31.5 percent annually. By 1992, Central America as a whole was exporting $500 million worth of nontraditional crops northward, mostly to the United States, each year.[71] Could this latest boom in farm production possibly lead to more equitable and less environmentally damaging gains?

Compared with previous booms in cotton and cattle, the current spread of nontraditional production seems attractive, both for satisfying the commercial needs of small peasant farmers and for protecting the environment. Most of the nontraditional crops in question can be produced in highland areas where poor peasants live; unlike cattle or cotton, they are suited to small plots and can be rotated in an environmentally sound fashion; they also provide more opportunities to employ pickers and packers (including women); and unlike some other regional crops, such as coffee, they mature quickly enough to encourage poor farmers to take a chance with their production. Thus, they can be a labor-using, land-saving form of farm intensification capable of generating wealth, even for small farmers, at low environmental cost.

Small peasant cultivators may nonetheless find it difficult to share in this new cropping boom, because the largest market shares are captured by farmers who gain access to credit, transportation, technical assistance, and timely marketing information. Governments have traditionally done a poor job of targeting the rural poor with such support. Poor and marginalized farmers are at a competitive disadvantage because they lack knowledge of the constantly changing tastes and quality-control requirements (for example, pesticide residue requirements) of the international marketplace. The growers' associations that could provide this information are dominated, like so much else in Central America, by a narrowly based elite. Fresh produce grown for export must also move to ports and airports quickly, and road systems leading out of the Central American highlands, where poor farmers live, are badly inadequate. Already, efforts to speed delivery and reduce pesticide residues on nontraditional crops have led to a restructuring of production away from the original satellite system, in which poor farmers were key participants, toward a more controlled system of sourcing, dependent on larger and more sophisticated producers.[72]

The obvious policy requirement is for governments in Central America to redirect their agricultural investments into poor areas, providing extension, road repair, and marketing assistance.[73] If this latest cropping boom is once again captured only by a narrow elite, rural social and environmental problems will continue to compound.

SUMMARY PRESCRIPTIONS

■ THE GREATEST THREAT TO THE RURAL environment in Latin America is the tradition of agricultural dualism and rural social inequity. Privileged commercial farmers who own most of the good land frequently mismanage the use of machinery or inputs, often because of government subsidies, with results that include excessive use of water, chemical pollution, and pesticide poisonings. Meanwhile, a majority of farmers are confined to poor lands they do not securely control, have difficulty protecting themselves from local elites and are uncertain of their ability to share in commercial gains from technical change. These disadvantages make them poor stewards of the land; they make few investments in resource protection and often farm were they can, in a "hit and run" fashion, cutting trees and plowing fragile lands as necessary. In Latin America, because of this extreme dualism in rural areas, first-generation and second-generation varieties of environmental damage are taking place side-by-side, at the same time.

To contain the second-generation damage, further reductions must be made in the subsidies that encourage privileged commercial farmers to use water and chemicals excessively. Subsidies that encourage ranchers to convert forest to pasture must also be phased out (Brazil, earlier the worst offender, has recently taken positive steps in this regard). To contain the first-generation damage inflicted by poor farmers, governments must go beyond technocratic remedies to consider substantial political and social reforms. Significant steps to redistribute land and make tenure secure would be the greatest help of all. Until the dispossessed rural majority become stakeholders in the environment and have a sense of secure local control over productive resources, farm-level investments in protecting the soil and forest will be few, and the current pattern of resource degradation will continue.

Part V
The Role of External Actors

INTERNATIONAL ASSISTANCE

■ AS THE PRECEDING SECTIONS HAVE SHOWN, a large variety of policy actions are needed at the local or national level. In some cases, national governments should do more (invest more in farm-to-market roads or undertake land reform, integrated pest management, or agricultural extension), and in other cases, governments should do less (cut subsidies for pesticides and cattle ranching, reduce trade protection, halt environmentally damaging rural mega-projects). The question is what the international assistance community can do to help, especially where local and national actions are slow or fail to materialize.

Traditional international assistance agencies, including both bilateral aid donors and multilateral lending institutions, have important roles to play in supporting appropriate actions at the national level. For bilateral donors, such as USAID, one of the most critical roles is to extend continued support for international agricultural research. For international financial institutions, money and technical transfers to support infrastructural investments and to sustain changes in policy are critical. These assistance functions have not been performed adequately of late, and too large a burden has fallen, by default, to NGOs. Until donors and international financial institutions do a better job of supporting environmentally sustainable gains in farm production in Asia, Africa, and Latin America, NGOs will remain overburdened and national governments will have a continued excuse for their own shortcomings.

U.S. SUPPORT FOR AGRICULTURAL RESEARCH

■ SCIENCE-BASED INNOVATION WILL BE critical to environmental protection in agriculture in the years ahead, both in the low-potential regions of Asia and Africa, where farmers will have to boost yields on a resource base already strained and degraded, and also in the high-potential regions of Asia, where high yields on overworked rice bowl and wheat belt lands have become more difficult to maintain without new knowledge.

Pierre Crosson and Jock R. Anderson have shown that science-based innovation is a cost-effective way to save land and water in farming, because the supply elasticity of new knowledge is greater than that of either land or water. More than 60 percent of all agricultural research programs in Asia have enjoyed spectacular annual rates of return above 50 percent, and even in Africa and Latin America, more than 40 percent of such programs have enjoyed rates of return above 50 percent. In one well-studied case, a research program at the International Institute for Tropical Agriculture developed a biological (rather than chemical) control to fight the cassava mealybug, with average benefit-to-cost ratios estimated, conservatively, at roughly 150 to 1.[74]

Modern agricultural research, a key to making environmental protection in farming cost-effective, does not happen automatically. Public support for agricultural research (both on-farm and off-farm) tends to be low in much of Asia and Africa, in part because of the economies of scale involved. It is hard for a small African country to develop its own critical mass of scientific talent and supporting infrastructure.[75] Private sector investment in farm research also tends to lag in the tropics, because servicing the needs of rich farmers in the temperate zone is commercially more attractive than servicing those of poor farmers in the tropics. Private sector investments in research, because of their commercial terms and motivation, are less likely in any case to respond to the special farming needs of the poor (a major concern in the area of biotechnology research, where private sector investment dominates).

Externally assisted public sector research is thus essential in Asia, Africa, and Latin America to assure the continued development of science-based knowledge of value to ordinary peasant farmers. U.S. foreign assistance has recognized this in the past and played a lead role, especially through funding contributions to the CGIAR, a system of 18 agricultural research centers located around the world, mostly in tropical countries. Since its formal creation in 1972, this unique system has generated research products of wide application and undeniable value. Plant breeding has been its traditional strength. By the 1980s, CGIAR germplasm was contained in more than 80 percent of all spring bread wheat grown in the tropics, 72 percent of all rice varieties, and more than 75 percent of maize varieties. Overall, some 80 percent of the calories and protein required in tropical countries derives from commodities whose productivity has been enhanced by CGIAR-generated agricultural technology.[76]

CGIAR research is a wise investment from an environmental standpoint. The highly responsive seed varieties traditionally developed by the CGIAR centers have given farmers important options to conserve land by increasing yield, and now the centers are specializing in innovations designed to benefit the environment directly, such as agroforestry, biological pest control programs (as in the cassava mealybug example noted above), and host-plant-resistance breeding strategies, which reduce the need for pesticides. The CGIAR centers help conserve germplasm by maintaining some 600,000 samples in gene banks.

This important system received adequate public funding in the past, thanks to foreign assistance. About two-thirds of all core funding traditionally came from bilateral foreign assistance, with 25 percent (about $40 to $48 million a year) coming from USAID alone. Given the relatively small size of this commitment, and given the large and well-established payoffs, we might expect CGIAR support to be relatively safe from the budget cutbacks taking place within USAID. Yet aid to CGIAR was slashed in the most recent round of these cuts. Core contributions from the United States were cut in terms of real dollars from $48 million in FY1992 to just $28 million in FY1994, with still more cuts expected for FY1995. The U.S. share of total international core contributions to CGIAR has declined from 25 percent to just 13 percent.[77]

Largely because of these weakened U.S. contributions, core funding to the CGIAR system as a whole has fallen in real terms more than 20 percent. As of late 1993, such funding cuts had already forced the research centers to release 110 internationally recruited scientists, plus more than 2,000 local employees. Roughly 20 percent of senior staff positions are now being abolished or left vacant, and the overall training effort of the CGIAR centers has been reduced roughly 45 percent.[78] Numerous important programs have been eliminated, including those covering mangrove rice in West Africa, production systems for wheat, virology work for potato, trypanosomiasis, and rice ecosystems.

Beyond the specific research programs and the relatively small dollar amounts at stake, U.S. support for international agricultural research has an important symbolic value. In part because the dollar figures are so small (compared with the large proven benefits), the continued abdication of U.S. leadership in this area sends an especially loud and unfortunate signal regarding the value attached by the United States to science-based agriculture. Until this signal can be corrected, the U.S. government will have little standing from which to criticize the performance of other governments in the area of environmentally sustainable farm production.

Respected U.S. authorities, from both the agriculturalist and environmentalist camps, have registered their dismay and puzzlement over this recent withdrawal of U.S. support for international research. In March 1994, testifying before the Agriculture Committee, U.S. House of Representatives, Dr. Norman Borlaug (winner of the 1970 Nobel Peace Prize for his work on high-yield food crops) said that

> the paradox facing all of us in research and extension today is that just when the need for greater effort in agricultural development has become so urgent, support to this work is being gutted.[79]

The end of the Cold War should have provided an important opportunity to revitalize U.S. bilateral assistance programs, especially in areas such as broadly based economic growth, democratization, and environmental protection. Instead, diminished threats to U.S. security have been used as an excuse to downsize many foreign aid programs. President Clinton's 1993 National Security Council study on foreign aid options stated flatly that, "with the disappearance of communism in the Soviet Union and Eastern Europe, the bedrock support for foreign assistance has eroded significantly. There is no clear vision guiding the shape of our foreign assistance agenda for a world without the U.S.S.R."[80] International research to promote environmentally sustainable agriculture is becoming one of the first victims of this post-Cold War loss of vision inside the U.S. development assistance community.

WORLD BANK SUPPORT FOR SUSTAINABLE AGRICULTURE

■ ALTHOUGH BILATERAL GRANT-AID PROGRAMS are well suited to less expensive actions such as funding for research, institution-building, and human resource development, the more expensive bricks-and-mortar assistance needed in agriculture—for irrigation and drainage or for farm-to-market roads—is typically provided through multilateral lending institutions such as the World Bank. The World Bank alone provides about 30 percent of all agricultural assistance to the low-income countries of Asia, Africa, and Latin America, administering a large portfolio of more than 400 agricultural projects representing $26 billion in loan value and $60 billion in total investments.[81]

Such investments will be essential in the future, especially if regions such as Africa and semi-arid Asia are to produce more food without further degrading the environment. Yet these kinds of investments have recently lost priority inside the World Bank. World Bank lending for all agricultural and rural development projects has been declining in real terms for more than a dozen years and is now roughly 20 percent lower than it was in 1978. Such lending was about 30 percent of total World Bank lending in 1978; by 1993, it was only 14 percent.[82]

These cutbacks occurred for three reasons. First was a widespread judgment, reached early in the 1980s, that falling international commodity prices reduced the economic justification for making loans to agriculture. Second was a decision, also taken early in the 1980s, to replace some investment lending in all sectors with lending for policy reform (the World Bank began stressing macroeconomic policy reform, privatization, and deregulation and made loans to secure such reforms under programs called structural adjustment). Third was a perceived risk that agricultural project lending by the World Bank would attract too much criticism from powerful environmentalists. None of these reasons, on reflection, justifies the sharp cut in World Bank lending for agriculture that has actually taken place.

International farm commodity prices are one standard that World Bank economists must use (to maintain their reputation for being market-driven and hard-headed) in judging the urgency of lending for agriculture; when those prices fell during the 1980s, a number of World Bank projects in the farming sector automatically were rated as less profitable and hence less worthy of support. Yet there is only a weak connection between international commodity prices and the real needs of farmers, or the protection of resources, in remote rural areas.

International commodity prices (which are primarily set in markets dominated or influenced by the farmers, consumers, and governments of rich countries) were indeed falling in the 1980s, but this did not mean that environmentally sustainable food production in the remote rural regions of Africa or Asia was suddenly less urgent. If anything, the opposite was true. The lower international commodity prices and market surpluses of the 1980s reflected not a sudden increase in food production in most of Asia, Africa, or Latin America so much as reduced growth in consumption, brought on by a debt crisis and a world recession (triggered in part by a U.S. switch to highly deflationary monetary policies). Food production increased rapidly in the 1980s, but mostly in industrial countries with large farm subsidies and already rich diets—including European countries and the United States—rather than in the places where hunger was an acute problem, such as Africa, semi-arid Asia, or Central America.[83] It would have made humanitarian sense for the World Bank to increase—not cut—its assistance to poor farmers in Asia, Africa, and Latin America at this point.

As for the World Bank's shift toward policy-based structural adjustment lending in the 1980s, this could have been a healthy move for agriculture and for environmental protection in agriculture, if it had brought an end to distorting input subsidies, devalued currency exchange rates adequately, decisively weakened government controls on marketing, and sufficiently improved terms of trade for farmers in relation to the industrial sector. As we saw in Africa, however, structural adjustment lending has not always brought about this set of desirable outcomes. Many borrowing countries have continued moving in and out of compliance with adjustment, and they have cut spending in the wrong place (less spending for research and farm extension, but continued spending for civil service salaries). By placing so much emphasis on public sector spending limits, the World Bank, together with its sister institution, the IMF, actually caused some governments, especially in Africa and Latin America, to reduce important investments in sustainable farming in the 1980s.[84]

Yet it is the World Bank's third principal motivation for cutting back on lending to the farm sector—its fear of criticism from environmentalists—that concerns us most. It is an irony that environmentalists have helped push the World Bank to do less in the agricultural sector, rather than obliging it to do more and to do it better.

Criticism of the World Bank by environmentalists has frequently been justified. Most World Bank projects, up through the mid-1980s, were undertaken with little attention to environmental costs or benefits. Because some of these projects, especially in the forestry and energy sectors, proved highly damaging to the rural environment (and also to forest and rural-dwelling people), influential environmental advocates, especially in the United States, became loud critics. The U.S. Congress, in 1985, passed legislation making further U.S. contributions to the World Bank contingent on environmental reforms. Inside the World Bank, lending officers suddenly faced a new burden of proof: They had to guarantee that a farm project involving road construction, irrigation, or chemical use (which, in Africa, are all badly needed) would not become a target for environmentalist critics. They found it safer, perhaps, to continue lending for structural adjustment or to make loans in some other sector.

Beginning in 1987, the World Bank took steps to try to appease environmentalist critics. A new central environmental department was created, with four regional units; all loans had to be screened for their environmental impact; borrowing countries were required to prepare comprehensive environmental action plans in order to be eligible for the World Bank's more concessional International Development Association assistance programs; a Vice-Presidency for Environmentally Sustainable Development was created in 1993; and later that year new policies and procedures were approved for releasing project information and for listening to appeals from citizens and NGOs regarding suspicious projects.[85] The World Bank also acknowledged that projects should be measured by their development impact, not just their rate of return.[86]

Changes were made in farm lending as well. The World Bank's new official lending strategy for the agricultural sector places less emphasis on structural adjustment lending and much more on lending for investment projects specifically intended to be of direct environmental benefit (such as natural resource monitoring, watershed management and reforestation, micro-catchment land use planning, improved small-scale irrigation technologies and water management, soil conservation, minimum tillage, intercropping, and residue management). The World Bank is also learning the practical value of moving toward greater participation at the community level (an internal study in 1990 showed that in virtually every participatory project financed by the World Bank, participation contributed to project effectiveness).[87] In 1994 the World Bank laid out a new model for pest control that went well beyond research in integrated pest management to include work with farmers and NGOs, and an internal World Bank-wide learning group on participatory development made sound recommendations regarding more participation by end-users. Without increasing the volume of lending, however, these reforms will go to waste.

Many environmentalists remain suspicious, and the World Bank's fear of their criticism continues to inhibit the revival of project lending in the agricultural sector. Without more lending, even the greenest loan procedures and requirements will have inadequate impact. World Bank officials have repeatedly promised to reverse the decline in farm lending that has been under way since the 1980s, but so far without result.[88] Friends of agriculture and friends of the environment should both be working to hold the World Bank to this pledge. It will be a shame if the healthy changes under way go unrecognized by critics. The large lending resources of the World Bank need to be mobilized and harnessed—and redirected back toward agriculture—if the goal of environmentally sustainable food production, especially in the poorest parts of Asia and Africa, is ever to be attained.

NONGOVERNMENTAL ORGANIZATIONS AS ENVIRONMENTAL ADVOCATES

■ WE HAVE SEEN THAT GOVERNMENTS and international assistance institutions have faltered in the pursuit of environmentally sustainable food production. Where they have fallen short, should NGOs try to step in?

NGOs cannot achieve sustainable production of food through a thousand-points-of-light strategy, one project at a time. However, they can play an important role not only directly by supporting agricultural projects but also indirectly by using their power, as advocates, to change government and foreign assistance policies.

Local and international NGOs have a proven record for implementing successful grass-roots projects in the area of protecting rural environmental resources. On the outer islands of Indonesia, CARE International has run dryland farming projects that have successfully employed water and soil moisture conservation technologies to shift land use from monocropped and degrading cassava production to lush and sustainable perennial farming of bananas, cocoa, coconut, and coffee. In some cases, these shifts in land use have taken hold so quickly that CARE has been able to withdraw its own support from specific local conservation areas within just three years.[89]

Such outcomes are valuable and gratifying. Yet there are limits to what NGOs can accomplish by working through projects alone. Going beyond the urge to substitute for governments, NGOs should dedicate themselves, more often, to changing government policy and to making the institutions of government more accountable to local farm communities. If restricted to implementing projects, they probably will not be able to extend their services and financial resources to a large enough share of the rural population. They will find it even more difficult to provide essential public goods, such as roads or research.

Consider the case of Bangladesh, which has one of the largest and most diversified NGO sectors in the world. At least 12,000 separate NGOs in Bangladesh are giving and receiving financial and technical support. Some are quite small, but some employ thousands of people. As numerous and widespread as these groups may be, and as much as they deserve the support of sympathetic outsiders, they have not yet been able to reach more than about 10–20 percent of the poor people in the country or to channel more than about 5 percent of total external aid. The successful Grameen Bank of Bangladesh (a grass-roots credit system for the poor) is rightly celebrated, but its overall impact is nonetheless limited. It provides a significant share of lending to the poor, but only about one-tenth of 1 percent of all national credit. All the NGOs in Bangladesh together provide only six-tenths of 1 percent of total credit.[90]

Compare this limited reach in Bangladesh to what can be achieved through committed government action. In the difficult area of family planning, Bangladesh has managed, since 1970, to increase contraceptive prevalence from 3 to 35 percent and thus to reduce the total fertility rate roughly 30 percent (from 7.0 to 4.9). The key to this success was a determined *governmental* family planning effort, launched after 1971 in part with foreign assistance and led by the Ministry of Health. This program constructed, equipped, and staffed 2,716 separate clinics and trained 23,000 female outreach workers, who are now bringing birth control and health assistance to 40 percent of all women of childbearing years at least four times a year (no easy task in a nation of 66,000 separate villages).[91] This powerful nationwide effort exceeds what could have been done by NGOs alone, on a project by project basis.

Like family planning, problems of sustainable food production are never going to be solved one project at a time. NGOs have at times been more effective than governments at reaching the poor (for example, Sarvodaya's success in setting up rural child care centers in Sri Lanka), but a strategy centered on NGOs is no substitute for good and effective government. The economically successful states of East Asia did not develop their economies through grass-roots projects implemented by NGOs. The critical land reforms in rural areas in East Asia, the low rates of inflation, and the mobilization of savings and capital investments, were all made available, in large measure, by active and competent government leadership.[92] Government-to-government foreign aid in some cases also played a key role.

One role for NGOs is to speed the emergence of such competent and accountable government authorities and encourage wise government-assistance policies in those regions of Asia, Africa, and Latin America that are still struggling. Trying to substitute for government authority in rural areas risks inviting a further abdication of official responsibility; instead, NGOs should use their energy and credibility to prod government authorities into performing the public sector tasks that they alone are equipped to handle.

NGOs performing as advocates should not separate themselves entirely from projects, since technical experience and direct contacts with rural communities are their chief source of creativity and public credibility. Yet they should pick their projects with the larger purpose of having an impact on governments. The 1993 *Human Development Report* of the United Nations Development Programme makes this point nicely:

> In eradicating poverty and providing social services, NGOs are unlikely ever to play more than a complementary role. Much more significant is their ability to demonstrate participatory models that

governments might follow—and to keep pressure on governments, in both the North and the South . . . . Their importance lies more in making the point that poverty can be tackled rather than tackling it to any large extent.[93]

To illustrate how international and national NGO advocacy can play this role in the area of sustainable agriculture, consider the problem of unsafe use of pesticides. Government leadership is essential to the effective promotion of integrated pest management, as in the case of Indonesia noted above. NGOs have an important role to play in launching broadly based advocacy campaigns designed to show citizens and officials the environmental advantage of alternative pest control strategies and to shame governments into a more active adoption of programs such as integrated pest management. The steadily increasing availability of public information about government policy and performance in this area gives NGOs a point of purchase. The FAO in Rome periodically conducts surveys to determine which governments are promoting integrated pest management and which are not. The comparisons generated from such surveys could provide one means for NGOs to hold laggards to account.

Also important is a newly created survey system within FAO and UNEP that could make both governments of developing countries and international chemical companies more accountable for international trade in potentially dangerous pesticides. This new system of prior informed consent (PIC), which the international NGO community (led by groups such as the Pesticides Action Network) fought hard to establish, asks importing country governments to declare publicly which chemicals, from a common list, they are willing to have imported into their country and which they are not. Lists of chemicals that have been restricted for trade country-by-country are then made available through the International Registry of Potentially Toxic Chemicals in Geneva, with an understanding that private companies, in both exporting and importing countries, should honor the restrictions. The transparency of this system provides local and international NGOs with abundant opportunities not only to shame private companies that continue to sell a restricted chemical but also (and sometimes more important) to shame importing governments for not including certain dangerous chemicals on their own list. At present, roughly 120 governments are participating in the PIC process.[94]

By showing the environmental benefits of alternative strategies and encouraging governments to take effective action, NGOs could help transform government promises on the environment into substantial actions. The recent spread of national environmental action plans (NEAPs) in Africa provides another setting in which this strategy could be attempted. Since 1987, the World Bank and other donors have encouraged the governments of Africa to incorporate environmental considerations into their overall economic and social development policies and to do so in a fashion open to citizen participation. More than 30 countries now have national environmental action plans in the process of development, approval, or implementation. NGOs can use this process first as a means to push governments for written or spoken commitments and then as a means to hold them to account.[95]

Advocacy is also a promising tactic for NGOs within industrial countries, including the United States.[96] Political authorities in rich countries do not yet adequately appreciate the future food production needs of Asia and Africa and the associated environmental danger of pursuing those needs without larger public investments or new science-based technical innovations. Too many U.S. political leaders carry a vision of environmentally sustainable agriculture that is disconnected from looming food production requirements, not sufficiently region-specific, and distorted by an industrial-country vantage point.

Even some of the strongest and most visible U.S. environmental policy leaders are prone to misstate the problem of sustainable production of food. In his 1992 best-selling book, *Earth in the Balance,* Vice President Al Gore pays little attention to the looming food production imperative in Africa and Asia and consistently downplays the environmental risk of responding to food needs with traditional low-input farming technologies. Gore grudgingly credits the green revolution with vastly increasing food production, but instead of warning about the danger of reduced spending for research, he disparages science-based innovation as a dangerous Faustian bargain. Gore pointedly associates green revolution farming with environmentally destructive techniques, implying (wrongly) that environmental protection would have been easier without higher yields.

The agricultural chapter in Gore's book does not focus on how to sustain higher production in the decades ahead so as to keep pace with population growth. Nor does it address how to boost yields in Africa or how to make governments more accountable in Latin America. It ignores all of the more urgent environmental dangers from farming that we have considered in this essay, such as loss of forest from slash-and-burn farming, overgrazing, nutrient depletion, watershed destruction, salinization, waterlogging, nutrient containment, or even pesticide abuse. Instead, it focuses on a more speculative difficulty—loss of germplasm.[97]

Genetic erosion in farm fields may be a problem in the U.S. corn belt, and it may eventually become a first-rank problem in the irrigated rice belt of East and South Asia, but it is hardly the place to begin addressing environmental dangers from (and to) farming in Africa, in dryland or upland Asia, or in Latin America.[98] Political leaders in the United States, perhaps because they are so close to the second-generation environmental problems that tend to afflict rich farmers in temperate zone countries, too often project those problems, erroneously, onto poor dryland farmers in the tropics.

NGOs in the United States thus have an important educational role to play, one that obliges them to master and explain agricultural and environmental imperatives with more care and precision on the region-specific basis necessary to secure a sound consensus. This is a difficult task, but one that has recently been taken up by at least one broadly based coalition of environmental and agricultural NGOs, the Washington-based Committee on Agricultural Sustainability for Developing Countries, chaired by former Ambassador Robert O. Blake at the World Resources Institute.[99] This committee, unique for its proven capacity to bridge gaps between agriculturalists and environmentalists, is educating and prodding political leaders in Congress, the executive branch, USAID, and the World Bank to turn greater energy and attention to the rapidly building agricultural environmental crisis in Africa, Asia, and Latin America.

Part VI
Conclusion

RESOURCE ABUSE AS POWER ABUSE

■ ENVIRONMENTAL DAMAGE FROM FARMING takes so many forms and has so many technical remedies that generalized conclusions seem elusive. Even when stressing regional differences, we run the risk of overgeneralization. Environmental problems in agriculture frequently differ not just between and within regions, but also between and within individual countries, individual watersheds and river valleys, or even individual farms.

Yet, at least one general pattern emerges. Environmental problems from farming in Asia, Africa, and Latin America tend to grow out of unbalanced relations of political and social power between farmers, between farmers and nonfarmers, or between farmers and governmental authorities. When power is out of balance within any one of these critical relationships, resources tend to be abused. Resource abuse, in this sense, reflects the abuse of power.

This is most obvious in Latin America, where badly unbalanced power relations exist between different kinds of farmers. A minority of socially privileged and politically powerful commercial farmers and cattle ranchers use their influence to monopolize high-potential land and then to secure input and credit subsidies from their governments, which induce them to cut too many trees, waste too much water, or spray too many chemicals. Their disproportionate power allows them to remain immune from punishment for the resulting environmental consequences. Meanwhile, a poor majority of socially and politically weak farmers is left with no choice but to farm on erodible dry and sloping land or to invade forest margins—and to do so in an environmentally destructive hit-and-run fashion for fear of being further displaced.

Environmental damage can also come when an unequal power relationship develops between farmers as a whole and urban dwellers (plus rural nonfarmers). In Africa and Asia, this happens in two quite different ways. In Africa, where farmers are more numerous but politically weaker than urban dwellers, governments adopt urban-biased policies that make the intensification of farming more difficult, hence blocking one of the best strategies available for protecting marginal lands and forests from the plow, the match, and the machete. In the industrializing economies of East Asia, however, farmers are better organized and more strong politically than urban dwellers (much like farmers in Japan or in other industrial regions such as Europe or the United States). These powerful farmers pressure their governments to adopt rural-biased policies such as farm subsidies and trade protection (just as in Japan, Europe, or the United States) that encourage the use of too much water and too many chemicals.

The critical role that power relations play is evident even in the location of the resulting environmental damage. In Africa, where farmers are weaker than nonfarmers, greater damage is felt on the farm itself, where the depletion of soil nutrients, degradation of rangeland, and loss of tree cover can produce an immediate disadvantage for poor farmers dependent on the land. In the industrializing economies of East Asia, where farmers are politically strong, greater damage is usually felt off the farm. Nonfarmers rather than farmers suffer the most from downstream chemical pollution, heavy residues of pesticides on marketed fruits and vegetables, and water shortages caused by excessive irrigation. Not only do the powerful cause environmental damage; they also consistently impose it on the weak.

In Africa and Asia, even greater damage can occur when rural resources come under the management of governments that are not democratic or are otherwise not accountable to their own citizens (farmers and nonfarmers alike). Publicly owned forest resources, publicly run state farms, and publicly developed irrigation or hydroelectric projects easily become venues for large-scale rural environmental destruction (in the industrial world, the former Soviet Union was a state-owned economy notorious for this sort of damage). The nonaccountable (and usually urban-dwelling) politicians, generals, and patronage bureaucrats who so often control these activities do not have the same interest as local rural dwellers in the long-term conservation of resources. And the rural dwellers themselves, if they sense that their own long-term control over a resource base is being lost to a distant public authority, may lapse into overexploitation to get as much as possible from the resource before it is taken away.

In our political geography of sustainable agriculture, a deeper cross-regional generalization therefore emerges. Abuse by poor farmers using traditional technologies in the wrong place or at the wrong time (first-generation abuse) is frequently an outgrowth of the political weakness of those farmers with respect to other farmers, nonfarmers, or their own government. Farming communities with sufficient power to control their own resources are seldom prone to destructive practices. Dangers arise when privileged rural elites or nonaccountable governments either take away or threaten to take away local control. Incentives to conserve resources then disappear. As Alan Durning has argued, "Access to a resource without control over it is calamitous. Nothing incites people to deplete forests, soils, or water supplies faster than the fear they will soon lose access to them."[100] Nonaccountable government agencies, such as forestry departments, centralized irrigation bureaucracies, or parastatal marketing boards, are often the greatest single threat to secure local control by ordinary farmers.

The imbalance of power also helps explain the second-generation environmental damage being done by wealthy commercial farmers. These farmers, situated at the other end of the political spectrum, are powerful enough to demand trade protection and subsidies, which induce excessive or distorted use of inputs. Much of the resulting damage (chemical pollution, groundwater depletion) is felt off-farm, where these powerful farmers escape accountability by wielding influence over local legal and political systems.

Democratization is one answer to the problems of poorly distributed or nonaccountable power in poor countries, but not so easy an answer as the word alone implies. Too often, democratization is initiated and dominated by an *urban* middle class. It does not always ensure adequate power for peasant farmers living in small rural villages. Political and social change must extend to the countryside itself.

New and less abusive relations of power must evolve between landed and landless, between indigenous and nonindigenous peoples, between men and women, and also between the city and the countryside overall. The ultimate objective should be to strengthen the control that ordinary peasant farmers exercise over the base of resources on which they and their children depend.

Linking the pursuit of environmental protection to this analysis of power relations is nothing new. In the agricultural sector, this linkage has been acknowledged at one time or another by a full range of conventionally technocratic institutions, including the World Bank, the FAO, the World Resources Institute, and the June 1992 United Nations Conference on Environment and Development.[101] It is nonetheless a daunting conclusion to reach, because it implies a need for real political change, in addition to technical or policy change, in order to win the looming battle to produce still more food without destroying the environment.

Notes

[1] For a review of recent efforts that industrial countries have made to remedy the environmental problems associated with farming, see Organization for Economic Co-operation and Development, *Agricultural and Environmental Policy Integration: Recent Progress and New Directions* (Paris: OECD, 1993).

[2] See Per Pinstrup-Andersen, *World Food Trends and Future Food Security*, Food Policy Report (Washington, DC: International Food Policy Research Institute, March 1994), p. 1. Including rice in milled form, FAO expects the consumption of cereal to increase from 918 million tons in the late 1980s to 1,460 million tons in the year 2010. See FAO, *Agriculture Towards 2010* (Rome: FAO, 1993), Table 3.5, p. 67; Robert D. Havener, "Population Growth and Sustainable Agricultural Production: An Emerging Agenda," paper presented at the conference 1989 World Food Production, Rio de Janeiro, Brazil, 5–9 November 1989.

[3] Kevin M. Cleaver, *A Strategy to Develop Agriculture in Sub-Saharan Africa and a Focus for the World Bank*, World Bank Technical Paper 203 (Washington, DC: Africa Technical Department, World Bank, 1993), p. 27; Carter Brandon and Ramesh Ramankutty, *Toward an Environmental Strategy for Asia*, World Bank Discussion Paper 224 (Washington, DC: World Bank, 1993), pp. 118–19; Billie R. DeWalt, Susan Stonich, and Sarah L. Hamilton, "Population and Land Use Changes in Honduras: A Regional and Community Level View," paper presented at the workshop Population Change and Land Use in Developing Countries, National Academy of Sciences, Washington, DC, December 1991; Paul S. Teng, "Integrated Pest Management in Rice: An Analysis of the Status Quo with Recommendations for Action," Department of Plant Pathology, University of Hawaii, Honolulu, January 1990, mimeo.

[4] All dollars are U.S. dollars. This list of indicators is based on the following sources: Joachim von Braun, Raymond F. Hopkins, Detlev Puetz, and Rajul Pandya-Lorch, *Aid to Agriculture: Reversing the Decline*, Food Policy Report (Washington, DC: International Food Policy Research Institute, October 1993); Robert Paarlberg and Michael Lipton, "Changing Missions at the World Bank," *World Policy Journal*, Vol. 8, No. 3 (Summer 1991), pp. 475–98; Montague Yudelman, "Feeding 10 Billion People in 2050: The Key Role of the CGIAR's International Agricultural Research Centers," Action Group on Food Security, Committee on Agricultural Sustainability for Developing Countries, Washington, DC, 20 April 1994, Table 3; Idriss Jazairy, Mohiuddin Alamgir, and Theresa Panuccio, *The State of World Rural Poverty* (New York: International Fund for Agricultural Development and New York University Press, 1992), p. 217.

[5] Richard E. Bissell, former assistant administrator for the U.S. Agency for International Development, has described this lack of agreement between agriculturalists and environmentalists as a new kind of "Natural Resource War." Richard E. Bissell, "The Natural Resource Wars: Let Them Eat Trees," *Washington Quarterly*, Vol. 17, No. 1 (1993), pp. 149–65.

[6] Richard G. Grimshaw, Christopher J. Perry, and James Smyle, "Technical Considerations for Sustainable Agriculture," in Jitendra P. Srivastava and Harold Alderman (eds.), *Agricultural and Environmental Challenges: Proceedings of the Thirteenth Agricultural Sector Symposium* (Washington, DC: World Bank, 1993), pp. 17–33.

[7] See, for example, Yujiro Hayami and Vernon W. Ruttan, *Agricultural Development: An International Perspective* (Baltimore, MD: Johns Hopkins University Press, 1985).

[8] Prabhu L. Pingali and Mark W. Rosegrant, "Confronting the Environmental Consequences of the Green Revolution in Asia," paper presented at the 1993 American Agricultural Economics Association international pre-conference Post-Green Revolution Agricultural Development Strategies in the Third World: What Next?, International Food Policy Research Institute, Washington, DC, August 1993.

[9] In 1967, William and Paul Paddock wrote an influential book depicting famine in India as inevitable. The Paddocks argued that since India could not possibly be saved, food aid should be terminated sooner rather than later, so that assistance could be redirected to more viable recipients in triage fashion. William Paddock and Paul Paddock, *Famine, 1975! America's Decision: Who Will Survive?* (Boston: Little Brown, 1967).

[10] M. S. Swaminathan, quoted in International Center for Maize and Wheat Improvement, *CIMMYT in 1992: Poverty, the Environment, and Population Growth, the Way Forward* (Mexico City: CIMMYT, 1992). See also the letter from M. S. Swaminathan to *Environment*, Vol. 36, No. 3 (April 1994), pp. 3–4.

[11] Vandana Shiva, *The Violence of the Green Revolution* (Penang, Malaysia: Third World Network, 1991).

[12] P. S. Appu, *Alleviation of Poverty in India* (Cambridge, MA: Harvard Center for International Affairs, May 1980), pp. 11–19.

[13] Masood Ahmad and Gary P. Kutcher, *Irrigation Planning with Environmental Considerations: A Case Study of Pakistan's Indus Basin*, World Bank Technical Paper 166 (Washington, DC: World Bank, 1992).

[14] N. S. Jodha, *Rural Common Property Resources: A Growing Crisis*, Gatekeeper Series 24 (London: International Institute for Environment and Development, 1991).

[15] Malcolm Gillis, "Multinational Enterprises and Environmental and Resource Management Issues in the Indonesian Tropical Forest Sector," in Charles S. Pearson (ed.), *Multinational Corporations, Environment, and the Third World* (Durham, NC: Duke University Press, 1987), pp. 85–86.

[16] Paul Faeth, "Building the Case for Sustainable Agriculture: Policy Lessons from India, Chile, and the Philippines," *Environment*, Vol. 36, No. 1 (January/February 1994), pp. 16–20, 34–39. Also, see the letter from M. S. Swaminathan of the Centre for Research on Sustainable Agricultural and Rural Development, Madras, India, to *Environment*, op. cit., pp. 3–4.

[17] Teng, "Integrated Pest Management in Rice," op. cit.

[18] Food and Agriculture Organization of the United Nations, "Integrated Pest Management in Rice in Indonesia: Status After Three Crop Seasons" (Jakarta: FAO, May 1988).

[19] Robert L. Paarlberg, "Managing Pesticide Use in Developing Countries," in Peter M. Haas, Robert O. Keohane, and Marc A. Levy (eds.), *Institutions for the Earth: Sources of Effective International Environmental Protection* (Cambridge, MA: Massachusetts Institute of Technology Press, 1993), p. 331.

[20] Genetic uniformity in irrigated fields of rice does not have to imply erosion of the larger pool of genetic material available to plant breeders. Despite the destruction of species in Asia, plant breeders still have access to varieties from upland and dryland areas and to laboratory collections of plant genetic material, such as the one kept by CGIAR. In the case of rice, the CGIAR gene bank at the International Rice Research Institute is one of the world's largest,

with roughly 86,000 total accessions, more than half of which have been protected for medium- and long-term storage. See Garrison Wilkes, *Strategies for Sustaining Crop Germplasm Preservation, Enhancement, and Use*, Issues in Agriculture, No. 5 (Washington, DC: CGIAR, 1992).

[21] Some new hybrid rice varieties can offer a 10–15 percent yield advantage over open-pollinated varieties, but the costs of seed propagation are extremely high, and the IRRI has estimated that only about 10 percent of the world's total rice area will ever be suited to commercial cultivation of such hybrids. See the letter from Klaus Lampe, Director General of the IRRI, to Dennis Avery, Hudson Institute, 12 October 1993, p. 2. For a more optimistic assessment, see FAO, *Agriculture Towards 2010* (Rome: FAO, 1993), p. 146.

[22] Montague Yudelman, "Demand and Supply of Foodstuffs up to 2050 with Special Reference to Irrigation" (draft), pp. 75–76.

[23] Pingali and Rosegrant, "Confronting the Environmental Consequences of the Green Revolution in Asia"; Derek Byerlee and A. Siddiq, "Sources of Growth on Wheat Yields in Pakistan's Punjab 1965–2000: Is There a Sustainability Issue?" Economics Working Paper, Vol. 90, No. 4 (Mexico City: CIMMYT, 1990).

[24] The widely noted tendency for governments in industrializing countries to shift from taxing to subsidizing farmers is well illustrated in East Asia. Japan shifted from negative agricultural protection in the nineteenth century to positive protection by the early twentieth and then to especially high protection after the mid-1950s. Korea and Taiwan subsequently followed a similar path. Between 1955 and 1982, the weighted average of nominal grain and meat protection in Korea increased from negative 15 percent to positive 166 percent, and in Taiwan from negative 21 percent to positive 55 percent. See Kym Anderson and Yujiro Hayami, *The Political Economy of Agricultural Protection: East Asia in International Perspective* (Sydney: Allen and Unwin, 1986).

[25] Yu-hsien Yu, "Current Agricultural Problems and Countermeasures in the Republic of China," Council of Agriculture, China, February 1991, mimeo.

[26] U.S. Department of Agriculture, *Agricultural Outlook, August 1993*, AO-199 (Washington, DC: U.S. Department of Agriculture, 1993), p. 57.

[27] Lester R. Brown argues that "when China turns to world markets on an ongoing basis, its food scarcity will become the world's scarcity." See Lester R. Brown, "Feeding China," *Worldwatch*, Vol. 7, No. 1 (September/October 1994), p. 19.

[28] The real cost of importing food and feed remains in a long-term secular decline; the temperate zone food and feed exporting countries (such as the United States, the European Union, Canada, Australia, and Argentina) continue struggling with too much production capacity—given the productivity gains that have steadily accompanied technical change—rather than too little. See D. Gale Johnson, *World Agriculture in Disarray*, Second Edition (New York: St. Martin's Press, 1991). Alarmists (including Lester Brown for at least two decades now) have repeatedly described this export growth capacity as close to exhaustion and have repeatedly been proved wrong. For a sharp critique of Brown's track record, see Dennis T. Avery, "The Worldwatch Institute vs. the Real World," in Avery, *Global Food Progress 1991* (Indianapolis: Hudson Institute, 1991), pp. 74–85.

[29] Theodore Panayotou, *Green Markets: The Economics of Sustainable Development* (San Francisco: ICS Press for Harvard Institute for International Development and the International Center for Economic Growth, 1993). For a larger discussion of environmental policy options in developing Asia, see Carter Brandon and Ramesh Ramankutty, "Toward an Environ-

mental Strategy for Asia: A Summary of a World Bank Discussion Paper" (Washington, DC: World Bank, 1993).

[30] William K. Jaeger, *The Effects of Economic Policies on African Agriculture,* World Bank Discussion Paper (Washington, DC: Africa Technical Department, World Bank, 1992); Cleaver, *A Strategy to Develop Agriculture in Sub-Saharan Africa,* op. cit., pp. 24–25.

[31] For the continent as a whole, estimates of the contribution that gains in yield have made to gains in total production vary widely, partly because yield gains in Africa are highly sensitive to fluctuations in rainfall and partly because gains in production overall have been so small. Depending on the time period selected, increased crop yields may have contributed as little as 6 percent to the small average annual increase in farm output registered during the difficult decade of the 1980s. See "Fueling Sustainable Growth in African Agriculture" (Muscle Shoals, AL: International Fertilizer Development Center, 1992). Yet since the early 1960s, for the production of cereals specifically, gains in yield may have accounted for as much as half of total growth in production. World Bank, *World Development Report 1992: Development and the Environment* (New York: Oxford University Press, 1992), Table 7.1, p. 135.

[32] FAO of the United Nations, *The Conservation and Rehabilitation of African Lands: An International Scheme* (Rome: FAO, 1990); Ridley Nelson, *Dryland Management: The "Desertification" Problem,* World Bank Technical Paper 116 (Washington, DC: Technical Department, World Bank, 1990), p. iii; Henry Noel Houerou, *The Grazing Land Ecosystems of the African Sahel* (Berlin and London: Springer-Verlag, 1987).

[33] Cleaver, *A Strategy to Develop Agriculture in Sub-Saharan Africa,* op. cit., p. 26.

[34] World Resources Institute, *World Resources 1992–93* (New York: Oxford University Press, 1992), p. 119.

[35] Bruce A. Larson, *Fertilizers to Support Agricultural Development in Sub-Saharan Africa: What Is Needed and Why,* Center for Economic Policy Studies Discussion Paper 13 (Arlington, VA: Winrock International Institute for Agricultural Development, 1993), p. 8.

[36] Cleaver, *A Strategy to Develop Agriculture in Sub-Saharan Africa,* op. cit., p. 32.

[37] Even if the prevalence of HIV were to reach 20 percent, the population growth rate—at the end of a 25-year epidemic—might be only about 1 percent lower. Jean-Louis Lamboray and A. Edward Elmendorf, *Combatting AIDS and Other Sexually Transmitted Diseases in Africa,* World Bank Discussion Paper 181 (Washington, DC: Africa Technical Department, World Bank, 1992).

[38] Farm size tends to be small in part because of the difficulty of keeping fields free from the invasive weeds that grow rampant. Parasitic weeds such as *Striga* attack cereals and food legumes in the arid savanna zones, while perennial grasses force farmers to abandon prime lands in the moist savanna. Okezie Akobundu, "Weeds in Human Affairs in Sub-Saharan Africa: Implications for Sustainable Food Production," *Weed Technology,* Vol. 5, No. 3 (July–September 1991), pp. 680–90.

[39] Arthur J. Dommen, *Innovations in African Agriculture* (Boulder, CO: Westview Press, 1988).

[40] World Bank, *World Development Report* (New York: Oxford University Press, 1992). These shortages can be hard for Africa's (mostly women) farmers to manage, given the large number of seasonal tasks (such as preparing the soil, planting seeds, mulching, or weeding) that must be carefully timed to local patterns of rainfall.

[41] Michael Crowder, *Colonial West Africa* (London: Frank Cass, 1978).

[42] Land Tenure Center, "Security of Tenure in Africa: A Presentation to the Agency for International Development" (Madison, WI: Land Tenure Center, April 1990).

[43] Montague Yudelman, *Prospects for Agricultural Development in Sub-Saharan Africa*, Occasional Paper (Morrilton, AR: Winrock Institute for Agricultural Development, 1987).

[44] World Bank, *Adjustment in Africa: Reforms, Results, and the Road Ahead* (New York: Oxford University Press, 1994), pp. 1–2, pp. 76–88.

[45] United Nations Development Programme, *Human Development Report 1993* (New York: Oxford University Press, 1993), p. 70.

[46] World Bank, "Kenya: The Role of Women in Economic Development," World Bank Country Study (Washington, DC: World Bank, 1989), pp. 9–24.

[47] As of July 1991, these four were Benin, Cape Verde, Comoros, Mauritius, and Namibia, with a combined population of just 8.4 million, or less than 2 percent of the total for Sub-Saharan Africa. See Jennifer A. Widner, *The Rise of a Party-State in Kenya* (Berkeley, CA: University of California Press, 1992).

[48] Lawrence H. Summers, "The Challenges of Development: Some Lessons of History for Sub-Saharan Africa," remarks prepared for delivery at the NIIA, Lagos, Nigeria, 8 November 1991; John Darnton, "Crisis-Torn Africa Becomes Continent of Refugees," *The New York Times*, 23 May 1994, p. A3.

[49] Overseas Development Institute, *Environmental Change and Dryland Management in Machakos District, Kenya: 1930–90*, Working Paper 56 (London: Overseas Development Institute, 1991), p. 38.

[50] Kenya's population is expected to increase from its current level of 25 million up to 83 million by 2025, and perhaps to 136 million by 2050. In some parts of western Kenya today, population densities are already above 1,000 persons per square kilometer, and typical farm sizes are down to less than 0.5 hectare of arable land. Farms of this size cannot produce enough of the fodder, fuelwood, and soil nutrients needed to support an average family, and the result has been the spread of resource abuse on common lands or on unsettled low-potential lands. Animals are turned lose to graze, soils are mined, and fuelwood is gathered through damaging raids on unprotected forest margins.

[51] Tom Forrest, *Politics and Economic Development in Nigeria* (Boulder, CO: Westview Press, 1993).

[52] Joyotee Smith, Anthony D. Barau, Abraham Goldman, and James H. Mereck, "The Role of Technology in Agricultural Intensification: The Evolution of Maize Production in the Northern Guinea Savanna of Nigeria," International Institute of Tropical Agriculture, Ibadan, 1993, mimeo.

[53] This is not for lack of warning. For example, see Robert J. Berg and Jennifer Seymour Whitaker (eds.), *Strategies for African Development* (Berkeley, CA: University of California Press, 1986).

[54] Cleaver, *A Strategy to Develop Agriculture in Sub-Saharan Africa*, op. cit., pp. 106–07.

[55] Graham Woodgate, "Agroecological Possibilities and Organizational Limits: Some Initial Impressions from a Mexican Case Study," in David Goodman and Michael Redclift (eds.), *Environment and Development in Latin America* (Manchester, England: Manchester University Press, 1991).

[56] Merilee S. Grindle, *State and Countryside: Development Policy and Agrarian Politics in Latin America* (Baltimore, MD: Johns Hopkins University Press, 1986).

[57] Alain de Janvry, *The Agrarian Question and Reformism in Latin America* (Baltimore, MD: Johns Hopkins University Press, 1981).

[58] Philip L. Martin, *Trade and Migration: NAFTA and Agriculture* (Washington, DC: Institute for International Economics, October 1993), p. 100.

[59] U.S. Department of Agriculture, "Environment and Food Safety Are Issues in U.S.-Mexican Trade," *Agricultural Outlook* (May 1992); Angus Wright, *The Death of Ramón González: The Modern Agricultural Dilemma* (Austin: University of Texas Press, 1990).

[60] Bonni van Blarcom, Odin Knudsen, and John Nash, *The Reform of Public Expenditures for Agriculture*, World Bank Discussion Paper 216 (Washington, DC: World Bank, 1993), p. 63.

[61] It must be hoped that these larger sales in the U.S. market, where consumers demand protection from the risks of pesticide residues, will encourage Mexico to continue upgrading its own rules governing the registration of pesticides and tolerance of residue. Between 1988 and 1991, in anticipation of NAFTA, Mexico reduced—from 35 to 19—the number of pesticides registered for use in Mexico, for which there was no residue tolerance in the United States. See Office of Technology Assessment, *U.S.-Mexico Trade: Pulling Together or Pulling Apart?* (Washington, DC: U.S. Congress, October 1992).

[62] See "Excerpt from U.S. Cable on Mexican Agriculture," *Inside NAFTA*, Vol. 1, No. 8 (20 April 1994), pp. 17–18. NAFTA could worsen the exodus by opening the Mexican market to cheap imports of U.S. corn. According to one worst-case estimate, as many as 668,000 workers (11 percent of the rural labor force) could leave the countryside (going either to urban areas or directly to the United States) if the *ejido* reforms proceed in the context of lower corn prices. See Sherman Robinson, Mary E. Burfisher, Raúl Hinojosa-Ojeda, and Karen E. Thierfelder, "Agricultural Policies and Migration in a U.S.-Mexico Free Trade Area: A Computable General Equilibrium Analysis," *Journal of Policy Modeling*, Vol. 15, Nos. 5–6 (1973), p. 695. See also Martin, *Trade and Migration*, op. cit., p. 108.

[63] Only about one-fourth to one-third of the Regional Solidarity Funds handled by Mexico's National Indigenous Institute have actually come under the control of indigenous organizations. See Jonathan Fox, "The Difficult Transition from Clientelism to Citizenship: Lessons from Mexico," *World Politics* 46 (January 1994), p. 173.

[64] In 1989, Mayan Indians in Chiapas formed their own coffee cooperative to export directly to the United States without having to sell (at low prices) to government-controlled marketing agents; the local municipal authorities reacted by putting several co-op members in jail and excluding their children from school. See "Revolt Helps Spawn Mexican Rights Movement," *The Boston Globe*, 8 March 1994, p. 10.

[65] Jonathan Fox, "The Roots of Chiapas," *Boston Review*, Vol. 19, No. 2 (April/May 1994), pp. 24–27; Samuel P. Huntington, *The Third Wave: Democratization in the Late Twentieth Century* (Norman, OK: University of Oklahoma Press, 1991). In Brazil's 1989 presidential race, the candidate of the political left took the large cities, while the candidate of the political right took the hinterland, where authoritarian clientelism was pervasive. In Mexico as well, the firmest base of electoral support for the long-ruling Institutional Revolutionary Party usually comes from rural areas, and in particular from the poorest rural areas, where traditional elites and party operatives have so much influence that they can control balloting outcomes.

[66] Robert G. Williams, *Export Agriculture and the Crisis in Central America* (Chapel Hill, NC: University of North Carolina Press, 1986).

[67] Sheldon Annis and contributors, *Poverty, Natural Resources, and Public Policy in Latin America*, U.S.-Third World Policy Perspectives No. 17 (New Brunswick, NJ: Transaction Publishers in cooperation with ODC, 1992).

[68] World Resources Institute, *Accounts Overdue: Natural Resource Depreciation in Costa Rica* (Washington, DC: World Resources Institute, 1991).

[69] DeWalt, Stonich, and Hamilton, "Population and Land Use Changes in Honduras," op. cit.

[70] John D. Strasma and Raphael Celis, "Land Taxation, the Poor, and Sustainable Development," in Annis and contributors, op. cit., pp. 143–69; Clive Bell, "Reforming Property Rights in Land and Tenancy," *World Bank Research Observer*, Vol. 5, No. 2 (July 1990), pp. 143–66.

[71] Stuart K. Tucker, "Equity and the Environment in the Production of Nontraditional Agricultural Exports," in Annis and contributors, op. cit., pp. 109–41. The boom in nontraditional crops extends beyond Central America. In 1993, Chile exported $1.5 billion worth of fresh fruit. In the northeast of Brazil, exports of tropical fruit are expected to increase fourfold between 1993 and 2000, reaching $200 million. See James Brooke, "Brazil Waters a Desert: The Desert Is Still Smiling," *The New York Times*, 25 March 1994, p. A4.

[72] Douglas L. Murray and Polly Hoppin, "Recurring Contradictions in Agrarian Development: Pesticide Problems in Caribbean Basin Nontraditional Agriculture," *World Development*, Vol. 20, No. 4 (1992), pp. 597–608.

[73] For an excellent summary of policy imperatives in the area of nontraditional exports, derived from the experience of Ecuador, see Lori Ann Thrupp, "Challenges in Latin America's Recent Agroexport Boom: Sustainability and Equity of Nontraditional Export Policies in Ecuador," *Issues in Development* (Washington, DC: World Resources Institute, February 1994), pp. 1–20.

[74] Pierre Crosson and Jock R. Anderson, *Resources and Global Food Prospects: Supply and Demand for Cereals to 2030*, World Bank Technical Paper 184 (Washington, DC: World Bank, 1992), p. 111; Hans Herren and Peter Neuenschwander, "Biological Control of Cassava Pests in Africa," *Annual Review of Entomology*, Vol. 36, 1991, pp. 257–283; R.B. Norgaard, "the Biological Control of Cassava Mealybug in Africa," *American Journal of Agricultural Economics*, Vol. 70, 1988, pp. 366–371; E. Evenson and Mark W. Rosegrant, "Determinants of Productivity Growth in Asian Agriculture: Past and Future," paper presented at the 1993 American Agricultural Economics Association international pre-conference "Post-Green Revolution Agricultural Development Strategies in the Third World: What Next?," Orlando, Florida, August 1993.

[75] Most governments in Asia, Africa, and Latin America tend to underinvest in agricultural research. See Jock R. Anderson, Robert W. Herdt, and Grant M. Scobie, *Science and Food: The CGIAR and Its Partners* (Washington, DC: World Bank, 1988), pp. 88–97.

[76] Letter from Lewis T. Preston, President of the World Bank; Jacques Diouf, Director General of the FAO; and James Gustave Speth, Administrator of the United Nations Development Programme, to Brian J. Atwood, Administrator of the U.S. Agency for International Development, 21 April 1994. Improved varieties of semi-dwarf wheat and rice developed at the CGIAR research centers have also enhanced the productivity of grain farmers in the United States. Even U.S. soybean producers, who complain loudly about research assistance to competitor countries such as Brazil, benefit from soybean breeding lines or varieties imported from laboratories in Brazil. Ralph W. Cummings, Jr., "Reverse Technology Flow as a Key to Future World and U.S. Agriculture," *BioScience*, Vol. 41, No. 11 (December 1991), pp. 775–78.

[77] Committee on Sustainable Agricultural for Developing Countries, "Annual Report 1993–94" (Washington, DC: World Resources Institute, 1994), p. 10. Percentage figures are from Yudelman, "Feeding 10 Billion People in 2050," op. cit., p. 10.

[78] With U.S. support waning, the World Bank is now trying to provide leadership. In May 1994, the chair of the CGIAR system at the World Bank, Ismail Serageldin, offered to provide additional funds up to $20 million to close roughly one-third of the anticipated funding gap in 1994 and 1995, providing that other donors, such as the U.S. government, would renew their own commitments. Despite this extraordinary appeal, USAID commitments to the CGIAR system remained paralyzed. *SAT News, Newsletter of the International Crops Research Institute for the Semi-Arid Tropics (ICRISAT)*, No. 15 (July–December 1993), p. 3; Committee on Agricultural Sustainability for Developing Countries, "Annual Report 1993–94," op. cit., p.13.

[79] See the testimony of Norman Borlaug before the Subcommittee on Foreign Assistance and Hunger, Committee on Agriculture, U.S. House of Representatives, 1 March 1994,p. 9. See also the testimony of Robert O. Blake, chairman of the Committee on Agricultural Sustainability for Developing Countries, before the Subcommittee on Foreign Operations, Committee on Appropriations, U.S. House of Representatives, 25 April 1994.

[80] Thomas W. Lippman, "Top U.S. Panel Urges Reform of Foreign Aid," *The Boston Globe*, 18 September 1993, p. 1.

[81] Ismail Serageldin, "Agriculture and Environmentally Sustainable Development," in Jitendra P. Srivastava and Harold Alderman (eds.), *Agriculture and Environmental Challenges* (Washington, DC: World Bank, 1993), p. 11.

[82] Michael Lipton and Robert Paarlberg, *The Role of the World Bank in Agricultural Development in the 1990s* (Washington, DC: International Food Policy Research Institute, 1990), p. 10. See also Hilary F. French, "Rebuilding the World Bank," in Lester R. Brown (ed.), *State of the World 1994* (New York: W. W. Norton and Co., 1994), pp. 167.

[83] FAO, *The State of Food and Agriculture 1991* (Rome: FAO, 1991), p. 31. See also Robert L. Paarlberg and Merilee S. Grindle, "Policy Reform and Reform Myopia: Agriculture in Developing Countries," *Food Policy* (October 1991), pp. 383–94; Robert L. Paarlberg, "Responding to Hunger Risks in the 1990s: The Responsibilities of Governments," statement to the Select Committee on Hunger, U.S. House of Representatives, 16 October 1991.

[84] Paarlberg and Grindle, "Policy Reform and Reform Myopia," op. cit.

[85] The World Bank's new independent Inspection Panel, empowered to receive and investigate complaints, officially began operating in September 1994.

[86] French, "Rebuilding the World Bank," op. cit., pp. 156–76.

[87] World Bank, "Agricultural Sector Review," Agriculture and Natural Resources Department, World Bank, Washington, DC, July 1993, pp. 22–23; World Bank, "Strengthening the Bank's Work on Popular Participation," External Affairs Department, World Bank, Washington, DC, 1990.

[88] Korinna Horta, an economist with the Environmental Defense Fund, wrote in 1993 that although the World Bank's reforms "have led to vastly increased environmental staff at the Bank and to several positive policy statements, such as the one strengthening the requirements for environmental impact assessments, there is mounting evidence that their impact in field projects has often been marginal at best." Korinna Horta, "Can the World Bank Be Reformed?" *EPA Journal* (April–June 1993), p. 26. World Bank leaders in 1993 promised $23 billion in new commitments to the sector over the next five years. See Serageldin, "Agriculture and Environmentally Sustainable Development," op. cit., p. 11.

[89] Grimshaw, Perry, and Smyle, "Technical Considerations for Sustainable Agriculture," op. cit., p. 31.

[90] Throughout Asia, Africa, and Latin America, NGOs channel only about 13 percent of net disbursements of official aid and only 2.5 percent of total resource flows. United Nations Development Programme, *Human Development Report 1993* (New York: Oxford University Press, 1993), pp. 92–95.

[91] World Bank, *Population and the World Bank: Implications from Eight Case Studies* (Washington, DC: Operations Evaluation Department, World Bank, 1992), pp. 39–41.

[92] Robert Wade, *Governing the Market: Economic Theory and the Role of Government in East Asian Industrialization* (Princeton, NJ: Princeton University Press, 1990).

[93] United Nations Development Programme, *Human Development Report 1993,* op. cit., p. 98. For another endorsement of this emphasis, by Idriss Jazairy, former President of the International Fund for Agricultural Development and current Executive Director of the London-based Agency for Co-operation and Research in Development, see Idriss Jazairy, "Sustainable Agriculture: The Role of NGOs," paper presented at the conference "How to Promote Sustainable Agriculture in Developing Countries," National Wildlife Federation, Washington, DC, 23–24 February 1994, p. 7.

[94] Robert Paarlberg, "Managing Pesticide Use in Developing Countries" and "Prior Informed Consent Update," Pesticide Action Network Updates Service, San Francisco, CA, posted 18 April 1994.

[95] François Falloux and Lee M. Talbot, *Crisis and Opportunity: Environment and Development in Africa* (London: Earthscan, 1993), p. 281. While advocacy-minded NGOs are strengthening the process of national environmental action plans in this fashion, that process, in turn, also helps to strengthen advocacy-minded NGOs. New organizations have sprung up precisely in those countries that have consistently adopted an open and transparent process of NEAPs.

[96] The International Development Conference is urging U.S. NGOs to adopt a "5 + 5" approach, allocating 5 percent of their program money to advocacy and 5 percent to development education. Something closer to a "10 + 10" approach might be in order in the area of sustainable agriculture, although any movement away from the direct delivery of services can be hard for NGO leaders to explain to individual contributors.

[97] According to Gore, "the single most serious strategic threat to the global food system is the threat of genetic erosion." See Albert Gore, *Earth in the Balance: Ecology and the Human Spirit* (New York: Plume, 1993), pp. 144, 321.

[98] Genetic uniformity in farming is not just a twentieth century danger. It helped bring on the devastating blight that caused the Irish potato famine of 1845 (all of the potato plants in Europe at the time were the descendants of two plants brought back from the New World). Modern science can, to an extent, protect against such dangers. When a somewhat similar southern corn leaf blight destroyed one-third of the U.S. corn crop in 1970, seed company scientists used their supplies of germplasm to breed a resistant variety of corn, which they provided to farmers the following year. See International Food Policy Research Institute, *1992 Facts and Figures: International Agricultural Research* (Washington, DC: International Food Policy Research Institute, 1992), p. 7.

[99] The committee is a network of 43 environment and development organizations, ranging from traditional agricultural groups, such as the National Association of State Universities and Land-Grant Colleges, to nonagricultural organizations, such as the Sierra Club and the National Wildlife Federation. For a review of the committee's recent activities, see Committee on Agricultural Sustainability for Developing Countries, "Annual Report, 1993–94" (Washington, DC: World Resources Institute, August 1993).

[100] Alan B. Durning, *Poverty and the Environment: Reversing the Downward Spiral,* Worldwatch Paper 92 (Washington, DC: Worldwatch Institute, November 1989), p. 42.

[101] World Bank, *World Development Report 1992,* op. cit., p. 143; FAO, "Elements for Strategies and Agenda for Action (Draft Proposal)," paper presented at the FAO/Netherlands conference "Agriculture and the Environment," 'S-Hertogenbosch, the Netherlands, 15–19 April 1991, p. 11; World Resources Institute, *World Resources 1992–93,* op. cit., p. 33–34; United Nations Conference on Environment and Development, *The Global Partnership for Environment and Development: A Guide to Agenda 21* (Geneva: UNCED, April 1992), p. 99.

About the Author

ROBERT L. PAARLBERG is Professor of Political Science at Wellesley College and Associate at the Harvard University Center for International Affairs. He writes widely on agricultural policy, environmental policy, and U.S. foreign economic policy. His most recent publications include *Leadership Abroad Begins at Home: U.S. Foreign Economic Policy After the Cold War* (Brookings Institution, 1994); *Assisting Sustainable Food Production: Apathy or Action?* (Winrock International, 1994); and "The Politics of Agricultural Resource Abuse," *Environment* (October 1994).

About the ODC

ODC fosters an understanding of how development relates to a much changed U.S. domestic and international policy agenda and helps shape the new course of global development cooperation.

ODC's programs focus on three main issues: the challenge of political and economic transitions and the reform of development assistance programs; the development dimensions of international responses to global problems; and the implications of development for U.S. economic security.

In pursuing these themes, ODC functions as:

■ *A center for policy analysis.* Bridging the worlds of ideas and actions, ODC translates the best academic research and analysis on selected issues of policy importance into information and recommendations for policymakers in the public and private sectors.

■ *A forum for the exchange of ideas.* ODC's conferences, seminars, workshops, and briefings bring together legislators, business executives, scholars, and representatives of international financial institutions and nongovernmental groups.

■ *A resource for public education.* Through its publications, meetings, testimony, lectures, and formal and informal networking, ODC makes timely, objective, nonpartisan information available to an audience that includes but reaches far beyond the Washington policymaking community.

ODC is a private, nonprofit organization funded by foundations, corporations, governments, and private individuals.

Stephen J. Friedman is the Chairman of the Overseas Development Council, and John W. Sewell is the Council's President.

Board of Directors